Particle Technology Forum

Held at the 2024 AIChE Annual Meeting

San Diego, California, USA
27-31 October 2024

ISBN: 979-8-3313-1675-4

Printed from e-media with permission by:

Curran Associates, Inc.
57 Morehouse Lane
Red Hook, NY 12571

Some format issues inherent in the e-media version may also appear in this print version.

Printed with permission by Curran Associates, Inc. (2025)

AIChE
120 Wall Street, FL 23
New York, NY 10005-4020

Phone: (800) 242-4363
Fax: (203) 775-5177

www.aiche.org

Additional copies of this publication are available from:

Curran Associates, Inc.
57 Morehouse Lane
Red Hook, NY 12571 USA
Phone: 845-758-0400
Fax: 845-758-2633
Email: curran@proceedings.com
Web: www.proceedings.com

TABLE OF CONTENTS

FLUIDIZATION: EXPERIMENTAL INVESTIGATION AND MODELING OF FLUIDIZATION PROCESSES

FLUIDIZATION: FUNDAMENTALS

MODELING FLUIDIZATION: APPROACHES, APPLICATIONS, AND FUNDAMENTAL INSIGHTS

NANOCOMPOSITES, COATINGS, AND HYBRID MULTISCALE SYSTEMS

NANOSTRUCTURED MATERIALS FOR BIOIMAGING AND THERANOSTICS

NANOSTRUCTURED MATERIALS FOR CATALYSIS AND ENERGY APPLICATIONS

NANO- AND MICROPARTICLES FOR PHARMACEUTICAL, BIOMEDICAL AND FOOD APPLICATIONS

PARTICLE BREAKAGE AND COMMINUTION PROCESSES: IN MEMORY OF THE LATE PROFESSOR LEONARD G. AUSTIN

PARTICLE CHARACTERIZATION

PARTICLE FORMATION PROCESSES

PARTICLE TECHNOLOGY FORUM AWARD PRESENTATIONS

PARTICLE TECHNOLOGY IN PRODUCT DESIGN AND MANUFACTURING

PARTICULATE MIXING AND SEGREGATION

PARTICULATE PROCESS MODELING AND PRODUCT DESIGN SESSION 1

PARTICULATE SYSTEMS: DYNAMICS AND MODELING: APPLICATIONS

PARTICULATE SYSTEMS: DYNAMICS AND MODELING: DISCRETE/CONTINUUM MODELS

PARTICULATE SYSTEMS: SOLIDS HANDLING, PROCESSING, CONVEYING, SEPARATION, AND HEAT TRANSFER

PHARMACEUTICAL POWDER AND PARTICULATE SYSTEMS

POSTER SESSION: PARTICLE TECHNOLOGY FORUM

POWDER AND PARTICULATE CHARACTERIZATION AND MEASUREMENT

POWDER HANDLING AND PROCESSING FOR BATTERY MATERIALS

PARTICULATE PROCESS MODELING AND PRODUCT DESIGN SESSION 2

PARTICLE TECHNOLOGY FOR PLASTICS RECYCLING, SURFACE MODIFICATION, ADDITIVE MANUFACTURING AND ENERGY STORAGE

HONORARY SESSION: REG DAVIES SPECIAL TOPICS (INVITED TALKS)

FLUIDIZATION: SUSTAINABILITY, RENEWABLES, AND DECARBONIZATION APPLICATIONS

FLUIDIZATION: COHESIVE MATERIALS AND OTHER INTERPARTICLE FORCES

INDUSTRIAL APPLICATIONS OF COMPUTATIONAL AND NUMERICAL APPROACHES TO PARTICLE FLOW

NANOSCALE MANUFACTURING: EXPERIMENTS, MODELING AND SIMULATION

147a A Review of Spent Catalyst Feeders in FCC Regenerators

Allan Issangya, Surya Karri, Francisco Sanchez Careaga

The fluidized bed regenerator is a key component of the fluid catalytic cracking (FCC) loop. Coke deposited on the spent catalyst particles in the regenerator is burnt off with air to restore the catalyst activity before it is reused in the riser reactor. Ideally, the spent catalyst returning from the stripper needs to be fed in the regenerator in such a way that provides a uniformly distributed catalyst and the incoming transport gas in the fluidized bed. Spreading the incoming spent catalyst uniformly in the regenerator improves solids mixing providing more efficient coke burning. Poor distribution In addition to the inefficient spent catalyst regeneration, there will be poorly fluidized regions of the bed that may lead to significant gas bypassing. Gas bypassing can cause "afterburn" in the freeboard which causes high temperatures in the freeboard which pose reliability concerns. A wide variety of solids distributor designs have been patented and many are in use in FCC regenerators. This paper presents a review of types of feed-solids distributors and their performance. Solids are introduced into fluidized beds as either dense streams via standpipes or as dilute feeds via risers. Although these solids distributors have been commercially used for decades, there is not much reported on the dispersion of the feed solids in the regenerator. To improve regenerator performance and meet the ever increasingly stricter environmental regulations there is need for more studies.

147b A Coarse-Grain CFD-DEM Approach Based on Particle Filtering for Simulating Fluidized Polydisperse Particles

Sathvik Bhat, Yuan Yao, Pedram Pakseresht, Yi Fan, Jorg Theuerkauf, Jesse Capecelatro

Highly resolved simulations of circulating fluidized bed (CFB) risers at industrial scale are prohibitively expensive. A common approach to reduce the computational cost is to lump particles into parcels, each containing many particles. However, existing approaches are based on ad-hoc corrections and fail to converge to the underlying deterministic equations in the limit the parcel approaches a single particle. In this work, a rigorous formulation of the filtered Computational Fluid Dynamics-Discrete Element Method (CFD-DEM) equations are presented. While exact, the equations result in unclosed terms, notably a sub-filtered drag force, that require new models. The unclosed terms are informed by highly resolved CFD-DEM simulations of moderately dense gas-solid flows. We consider simulations of monodisperse and polydisperse Geldart-A particles in unbounded flows. Parcels are identified from the highly resolved simulations efficiently, in a consistent manner with the underlying formulation, using a KDTree algorithm. Collisions between the parcels are handled using soft-sphere approach that modifies the coefficient of restitution based on the number of particles per parcel. Variation in particle size and velocity within each parcel is quantified. The relative contribution of the sub-filtered drag is found to increase with number of particles per parcel, and depend on the local filtered volume fraction and Reynolds number. Symbolic regression is employed to arrive at close-formed algebraic models. The accuracy and speed-up of the proposed filtered CFD-DEM approach compared to the highly resolved simulations will be presented.

147d Investigation of Heat Transfer in Fluidized Bed Heat Exchangers

Krutika Appaswamy, Aaron Morris, Zhiwen Ma

Fluidized bed heat exchangers play an important role in various industrial processes, where efficient heat transfer is required. In this study, we look at the heat transfer phenomena within fluidized bed heat exchangers using two different modeling approaches: high-fidelity discrete element modeling (DEM) and computationally inexpensive two-fluid modeling (TFM). Both models are validated against experimental data for heat transfer coefficients in a parallel-plate narrow channel fluidized bed heat exchanger. In TFM, since the solids are not individually tracked, the model is computationally inexpensive but requires closures for the constitutive relations that dictate the solid stresses and heat flux. To obtain the effective solid conductivity in TFM, a continuum conduction model developed by Morris et al. is implemented. The conduction model uses chute flow DEM data to obtain a Nusselt number correlation which is a function of solid concentration. Our findings show that DEM outperforms TFM in capturing heat transfer trends observed experimentally. To understand the disparities between the two models, we decouple heat transfer and hydrodynamic differences, examining these factors individually. Specifically, for studying the heat transfer, we validate the Nusselt number correlation, originally developed for chute flows, for its applicability in fluidized beds using DEM. Scaled Nusselt number is plotted against the solid fraction and the effect of particle roughness as well as conduction lens radius on the curve is studied. It is observed that these parameters have a significant effect on the wall to particle heat transfer. Hence, the Nusselt number correlations need to be modified to include the effects of these parameters. Furthermore, we validate the closures used in TFM for predicting heat transfer by applying them to DEM hydrodynamic flow fields. To enhance the accuracy of predictions, histograms are used to study the noise in the Nusselt number when plotted against solid fraction. Notably, curve fits to the means of these histograms prove to be more accurate predictors when applied to DEM flow fields. Additionally, we also observe consistent trends in the histograms for a given particle surface roughness across different solid fraction values. Upon plotting the standard deviation of each histogram against the solid concentration, we observe consistent curves across different fluidizing velocities for the same surface roughness. This consistency further emphasizes the importance of parameter considerations and accurate modeling techniques for reliable predictions in fluidized beds. Overall, our study contributes to a deeper understanding of heat transfer mechanisms in fluidized bed heat exchangers. Since we have validated our heat transfer models, in order to obtain a comprehensive method to improve TFM predictions, next steps would include studying the hydrodynamic predictions of the models.

147e Study on Process Intensification of Granular Flow of Using Pulsed Load

Haifeng Lu, Liang Zhang, Xiaolei Guo, Haifeng Liu

The transition of particle systems between solid-like and fluid-like regimes is driven by changes in density and in applied states. In this respect, pulse loads in drive the unjamming transition have been proved to reduce the unconfined yield stress and cohesion, thus achieving effective particle fluidization. In the present work, we have studied the effects of vibration, aeration, and pulse aeration on powder discharge. We investigated the dynamic behavior of powder discharge in a 2D powder bed under vibration, aeration, and pulse aeration, respectively. The effect of powder physical properties and modulated pulse parameters (frequency, duty cycle, etc.) on particle fluidization was analyzed. The dynamic characteristics of powder discharge and agglomerates motion and deformation were studied using digital image processing and signal analysis. The effects of the modulated pulse load in improving slugging, suppressing channeling, breaking arches were revealed. The mechanism of modulated pulse load in improving powder flowability was revealed.

147f Study on Evolution of Wet Agglomerates and Dispersion of Wet Particles in Gas-Solid Fluidized Bed

Congjing Ren, Jingyuan Sun, Jianan Cheng, Jingdai Wang

The process of fluidized bed with liquid spray and the direct connection process of a loop reactor and a fluidized bed are two original processes developed by Zhejiang University in cooperation with Sinopec Tianjin Petrochemical Corporation for the production of high-end polyethylene products. This paper carries out the research on the agglomeration evolution and dispersion characteristics of wet particles in the fluidized bed. In this paper, the gas-solid fluidized bed with liquid spray / slurry spray corresponding to the processes is taken as the research object. By means of cold-model experiment, mathematical modeling and hydrodynamic simulation, this paper explores the dynamic evolution mode and law of wet agglomerates, the dispersion characteristics of slurry particles and the assignment relationship of slurry particles in different zones of the fluidized bed.

147g Flow Transitions and Effective Properties in Multiphase Taylor–Couette Flow

Abhishek Shetty, Melany Hunt, Arthur Young

The properties of multiphase flows are challenging to measure, and yet effective properties are fundamental to modelling and predicting flow behaviour found throughout many geophysical and industrially relevant flows. In the present work, an air bearing rheometer coupled with a powder flow cell is used to carry out a careful exploration of a gas-fluidized bed in which the fluidization rate and the rotational speed can be varied independently. The measured torque displays a range of rheological states: quasistatic, dense granular flow behaviour at low fluidization rates and low-to-moderate shear rates; turbulent toroidal-vortex flow at high shear rates and moderate-to-high fluidization rates; and viscous-like behaviour with rate-dependent torque at high shear rates and low fluidization or at low shear rates and high fluidization [1]. To understand the solid-like to fluid-like transitions, additional experiments were performed in the same rheometer using single-phase liquid and liquid–solid suspensions. The results further demonstrate that these different multiphase flows undergo analogous flow transitions at similar Bingham or Reynolds numbers once effective properties are computed which allows one to compare the results across all the experiments. The corresponding dimensionless torques associated with these experiments also show comparable scaling in response to annular shear.

References

[1] Young, A. B., Shetty, A., & Hunt, M. L. (2024). Flow transitions and effective properties in multiphase Taylor–Couette flow. *Journal of Fluid Mechanics*, *983*, A14.

147h A Monte Carlo Approach to Modeling Fluidized Systems

Matthew Black, Aaron Morris

The ability to accurately and efficiently model fluidized systems is of great practical interest to researchers and engineers. Models such as the Multiphase Particle-in-Cell (MP-PIC) method have leveraged statistical properties and interpolation functions to simulate large-scale systems. An alternative statistically-driven approach is presented here, based on the Direct Simulation Monte Carlo (DSMC) method. Typically used to model rarefied gases, the DSMC method has been extended to solving the revised Enskog equation for dilute granular flows. By directly solving this equation, the DSMC method does not rely on solid stress closures that are needed for other continuum modeling approaches. Previous attempts to apply DSMC theory to granular flow systems have encountered difficulties with over-packing because the collisional pressures were not properly accounted for. In our previous work, overpacking in granular systems was prevented by the implementation of a random walk algorithm during time integration. This addition greatly improves the outcomes of a DSMC-based method when simulating dense to moderately dense flows. In this work, we assess the validity of the DSMC method across several fluidization regimes. The DSMC predictions are compared against those from DEM simulations, and metrics such as the bed pressure drop and defluidization curves are compared. It is demonstrated that the DSMC approach provides a reasonable approximation of the results from DEM in several applications for a lower computational cost. Additionally, areas where this method is currently deficient are identified and avenues of improvement for future work are explored.

85a Analysis of Solids Volume Fraction and Dimensionless Regime Mapping in a Circulating Fluidized Bed Riser

Steven Rowan, Ronald Breault, Justin Weber

One of the most difficult challenges faced when trying to apply process information obtained from fluidization and hydrodynamic testing in small scale circulating fluidized beds to larger pilot or commercial scale units is identifying the correct scaleup methodology. Historically, there have been several approaches that have met with varying levels of success, including similarity methods that rely on defining and matching the values of dimensionless numbers involving different length scales.

Recent efforts by Breault et al. have focused on developing a method utilizing a dimensionless regime map incorporating use of two dimensionless ratios: U_g/U_{tr2} – the ratio of the superficial gas velocity to the upper transport velocity, and Gs /Gs* - the ratio of the solids circulation rate to the saturation carrying capacity of the fluidizing gas. These recent efforts have led to the development of more accurate correlations for the Gs* as a function of riser diameter for both Geldart A- and Geldart B-type particles.

Utilizing these newly developed correlations for Gs*, fluidization data for 200-micron glass beads obtained from experiments previously conducted in the 12-inch (0.3m) diameter cold flow circulating fluidized bed (CFCFB) unit at the National Energy Technology Laboratory's (NETL) Morgantown, WV site have been revisited considering the recently developed dimensionless regime map. The operating conditions for these experiments, when placed on the dimensionless regime map, illustrate that the tests conducted covered a range of fluidization regimes, including the dilute upflow, core annular flow, and fast fluidization regimes. This paper presents a review of the resulting dimensionless regime map via statistical and chaos analysis of riser solids fraction profiles, including recurrence analysis, to illustrate the macroscopic behavior within these different fluidization regimes.

85b Effect of Viscosity on a Slip Velocity in a Liquid-Solid Circulating Fluidized Bed

Nirmala Gnanasundaram, Muruganandham Loganathan, Kumar Perumal, Chitra Devarajulu

This work aims to experimentally study the slip between the phases for a variable liquid viscosity in a 2m length, liquid solid circulating fluidized bed column. The effect of total liquid velocity and liquid viscosity on slip velocity was studied and found that slip velocity is greater than the particle terminal velocity and it decreased with an increase in secondary and viscosity of liquids for all the solids considered in the study. A generalized correlation to predict slip velocity in an LSCFB riser which depends on size, density of solids and viscosity of fluid has been developed. The predicted model was found to be validated with the experimental results. The study also covered the analysis of maximum liquid and solid velocity

85c The Advanced Scale up Reactor Experiment (ASURE) Facility: A Testbed for Demonstration of Advances in Biomass and Waste Co-Gasification Systems.

Steven Rowan, Ronald Breault

The Advanced Scale Up Reactor Experiment (ASURE) facility factors heavily in the National Energy Technology Laboratory's (NETL) Gasification Multi-Year Research Plan (MYRP) to support the U.S. Department of Energy's Office of Fossil Energy and Carbon Management's (FECM) stated goals to "advance and demonstrate modular biomass and waste gasification systems for carbon-neutral hydrogen production through cost reduction, co-gasification, and process intensification".

The ASURE reactor will provide a pilot scale fuel-flexible testbed for advancing the TRL of emerging technologies for co-gasification of carbonaceous feedstocks (including biomass, plastics, and other waste materials), catalytic production of hydrogen and other high value chemicals, ash and syngas tar cleanup, testing of refractory for use with high-alkali feeds, and generation of high-fidelity experimental data that can be used for computational model verification and validation.

The discussion presented in this paper details how the ASURE reactor will support FECM's stated goals for gasification-related research, as well as the design requirements that will lead into the planned conversion of the existing chemical looping reactor (CLR) unit located at NETL's Morgantown, WV site into a circulating fluidized bed (CFB) gasifier capable of operating at temperatures and pressures up to 1000°C and 11 bar, respectively. Additional discussion detailing the potential for external collaboration and use of the ASURE reactor for pilot scale testing of developing gasification related technologies will be included.

85d Descent of Intruders in Vibrated Gas Fluidized Beds

Oscar Punch, Qiang Guo, Maty Gueye, Javad Omidi, Christopher Boyce

We investigated the descent of intruder particles in pseudo-2D vibrated gas fluidized beds. Intruders descended with an oscillating velocity with the descent velocity varying with vertical position in the fluidized bed. We attribute the varying descent velocity to varying particle packing fraction at different locations in the fluidized bed. We also found that the intruder descent velocity decreases dramatically as the intruder is moved closer to a side wall. We also investigated the interaction of two intruders finding high levels of interaction, including a phenomenon similar to drafting-kissing-tumbling observed in Newtonian fluids.

85e Faraday Wave Patterns in Fluidized Granular Materials: Effects of Combined Vertical and Horizontal Vibration

Javad Omidi, Oscar Punch, Qiang Guo, Christopher Boyce

This study thoroughly explored the complex dynamics and determining the onset mechanisms of Faraday waves within a fluidized granular materials subjected to simultaneous vertical and horizontal vibrations, using two-fluid model (TFM) simulations. Through dimensional analysis, a non-dimensional empirical correlations and regime maps for both wavelength and wave height were established. Intriguingly, it is observed that an increase in vertical vibration frequency results in a reduction in horizontal mixing, while horizontal mixing is amplified by an augmentation in horizontal vibration frequency. Furthermore, it is found that the pace of vertical mixing is heightened and horizontal mixing rates are concurrently diminished by elevating gas flow rates.

85f Magnetic Resonance Velocimetry of Particle Hydrodynamics in a Three-Dimensional Draft Tube Spout-Fluid Bed

Jens P. Metzger, Boyuan Chen, Alexander Penn, Christian Guenthner, Klaas P. Pruessmann, Christoph R. Müller

Draft tube spout-fluid beds (DTSFB) are widely used industrial reactors for processes requiring intensive mixing of solids and high heat and mass transfer rates between contacting gas and solid phases, such as particle drying[1] and coating[2]. Due to the presence of a draft tube, the solid flow in a DTSFB is compartmentalized in a slow annular region and a fast spout in the draft tube, whereby the gap height of the draft tube regulates the particle circulation rate and the gas contacting time[3].

However, the inherent opacity of granular materials impedes the visual examination of the complex, internal hydrodynamics found in DTSFBs. Hence, semicircular beds[4] or visual probes[5] are often used to inspect the dynamics, which, in turn, perturb the particle flow due to the presence of additional walls or the insertion of probes.

The present work employs magnetic resonance imaging (MRI) to non-invasively study the hydrodynamics of the solid phase in a fully three-dimensional DTSFB[6]. To this end, a lab-scale spout-fluid bed of oil-filled particles is equipped with a draft tube and measured by a 16-channel radio-frequency detector array in a 3T MRI scanner[7,8]. Solid distribution maps and particle velocity maps are obtained for various operational settings of the DTSFB, allowing to assess the entrainment of solid material from the annular region to the draft tube and, additionally, revealing flow phenomena such as a vena contracta flow at the entrance of the draft tube and the suppression of gas bubbling in the annular region of the bed. Complementary simulations using the discrete element method coupled with computational fluid dynamics confirm the experimental observations and provide further insight into the gas flow field of the DTSFB which is currently inaccessible by magnetic resonance imaging.

References

1. Olazar M, Lopez G, Altzibar H, Amutio M, Bilbao J. Drying of Biomass in a Conical Spouted Bed with Different Types of Internal Devices. *Dry Technol.* 2012;30(2):207-216.
2. Plawsky JL, Littman H, Paccione JD. Design, simulation, and performance of a draft tube spout fluid bed coating system for aerogel particles. *Powder Technol.* 2010;199(2):131-138.
3. Grbavčić ŽB, Littman H, Morgan MH, Paccione JD. Spouted and spout-fluid beds with draft tubes. In: Grace JR, Epstein N, eds. *Spouted and Spout-Fluid Beds: Fundamentals and Applications.* Cambridge: Cambridge University Press; 2010:128-140.
4. Saadevandi BA, Turton R. Particle velocity and voidage profiles in a draft tube equipped spouted-fluidized bed coating device. *Chem Eng Commun.* 2004;191(10):1379-1400.
5. He Y-L, Qin S-Z, Lim CJ, Grace JR. Particle velocity profiles and solid flow patterns in spouted beds. *Can J Chem Eng.* 1994;72(4):561-568.
6. Metzger JP, Chen B, Penn A, Guenthner C, Pruessmann KP, Müller CR. Magnetic resonance velocimetry of particle hydrodynamics in a three-dimensional draft tube spout-fluid bed. *Chem Eng J.* 2024;485:149678.
7. Penn A, Tsuji T, Brunner DO, Boyce CM, Pruessmann KP, Muller CR. Real-time probing of granular dynamics with magnetic resonance. *Sci Adv.* 2017;3(9).
8. Penn A, Boyce CM, Pruessmann KP, Müller CR. Regimes of jetting and bubbling in a fluidized bed studied using real-time magnetic resonance imaging. *Chem Eng J.* 2020;383:123185.

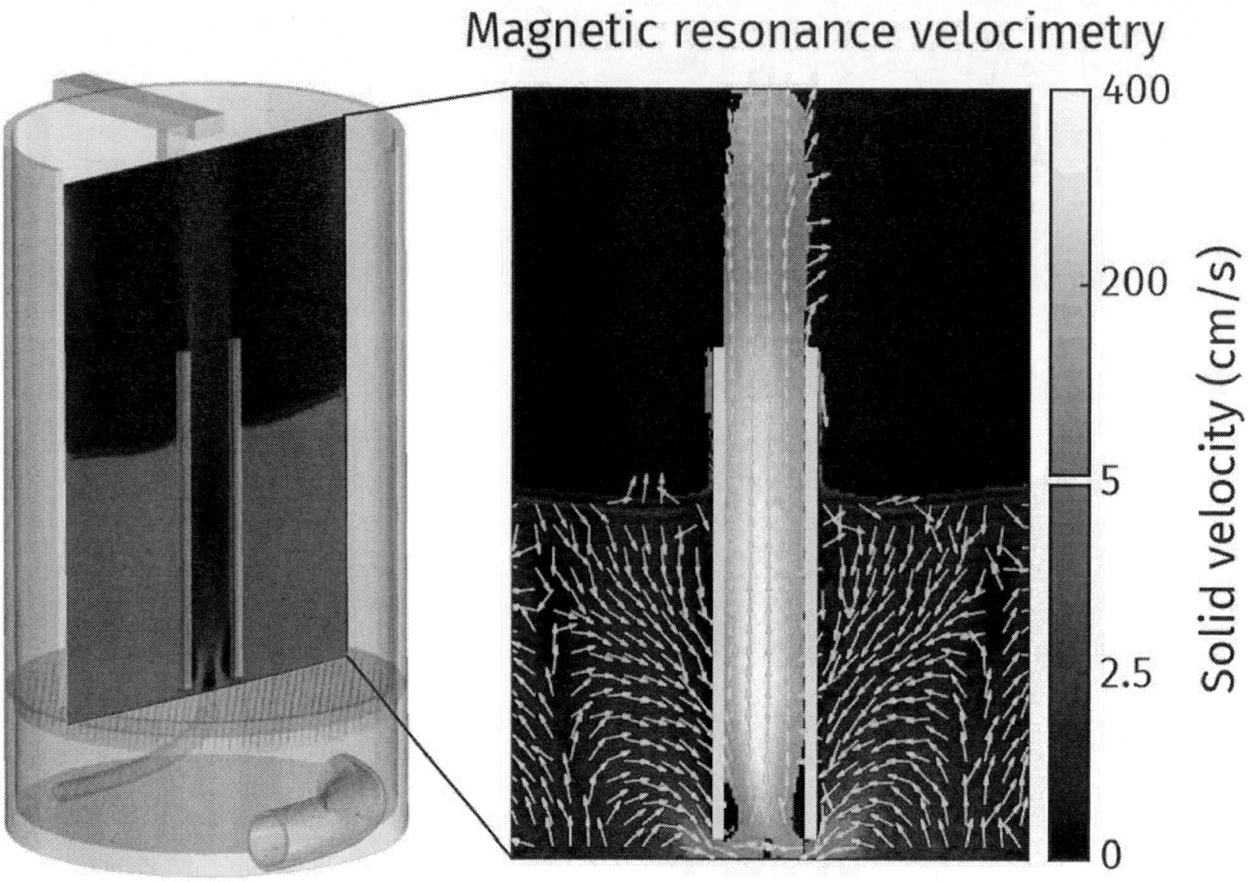

Magnetic resonance velocimetry
400
200
5
2.5
0
Solid velocity (cm/s)

215a Oscillated Gas-Injection in Granular Matter: Exploring Structured Bubbling Flow

Javad Omidi, Christopher Boyce

Understanding the dynamics of granular materials that resemble liquid flow is crucial in both natural and industrial processes. Industries like food, pharmaceuticals, and clean energy, use bubbling fluidized beds, where gas flows upward through particles suspended in a liquid-like state. Our study shows that oscillating gas flow at 7Hz can organize chaotic bubble motion, allowing precise control over particle and gas behavior. In this two-fluid model (TFM) simulation, after validating our computational model against experimental data, we divided the injection surface into various parts, ranging from 2 to 10 slices, with constant phase differences between consecutive slices. As we increase the phase difference between slices from $\pi/4$ to π, we observe an increase in the number of bubbles in the bed, and consequently, a decrease in both the size of the bubbles and the vertical distances between them. For a constant phase of $\pi/4$, as the number of slices increases from 4 to 10 at the bottom, the structure pattern modifies and the number of bubbles in the pattern increases, while the size of the bubbles decreases. Additionally, the results indicated that for an odd number of slices, the pattern tends to be more centered, whereas for an even number of slices, the pattern appears more evenly distributed across the horizontal direction. The scalable structure and the method introduced here offer solutions to key scale-up and operational challenges, enhancing the applicability of bubbling fluidized beds in both existing and emerging technologies.

215b A CFD-DEM Study to Quantify the Influence of Particle Clustering on Catalytic Reactions

Balivada Kusum Kumar, Himanshu Goyal

Gas-solid catalytic reactors like risers are essential in the chemical industries. The multiphase flow in these reactors leads to the formation of particle clusters, influencing the contact between the gas and particles and reducing the reactant conversion. Three length scales can be identified in these systems – particle size ~O(mm), cluster size ~O(cm), and reactor size ~O(m). Quantitative models that can quantify the impact of particle clusters on the catalytic reactions are limited. To this end, we perform Computational Fluid Dynamics – Discrete Element Model (CFD-DEM) simulations of particle clustering in a triply periodic domain. This simulation setup is representative of the fully developed region in a riser reactor. The simulations are performed for various Damköhler numbers by changing the reaction rate coefficient. We utilize our recently developed DBSCAN-based methodology to obtain the cluster properties, such as volume, area, and chord length. DBSCAN is a commonly used unsupervised machine learning (ML) algorithm. A distribution of cluster size is found, which is characterized by a probability density function (PDF). The effective cluster size that controls the conversion of gas-solid reaction is estimated. A reduced-order model is developed based on the effective characteristic length of the cluster. The predictions of the reduced order model are compared with those of the CFD-DEM simulations.

215c Elucidating Chromium Contaminates Distribution in a Circulating Fluidized Bed for Light Alkane Dehydrogenation

Yupeng Du, Yuchao Zhao

Introduction

Propane dehydrogenation (PDH) becomes more and more attractive to produce propylene. Commercial PDH plants employ either fixed beds (e.g. Catofin process) or moving beds (e.g. Oleflex process). However, both technologies are facing the problems of frequent regeneration of catalysts and/or reaction heat supply due to the highly endothermic nature of the dehydrogenation reaction and the fast deactivation of the catalysts. This has triggered the utilization of the fluidized bed technology for PDH to help alleviate those shortcomings, where the chromium-based catalyst is commonly used. Although it is well-known that Cr^{6+} is quite harmful to the environment and human health, its distribution in a Circulating Fluidized Bed (CFB) setup is still unclear.

Methodology

The distributions of chromium with various valences (i.e., Cr^{6+}, Cr^{3+}, and Cr^{2+}) on the dehydrogenation catalysts, the detailed fluid flow, heat and mass transfers, and coupled chemical conversions in a whole CFB unit need to be precisely quantified. Fortunately, the Multi-Phase Particle-In-Cell (MP-PIC) methodology, has offered real possibilities to vary many physical parameters linked to CFBs and study their effects on the process performance at large scales. Thus, MP-PIC simulations on the whole CFB unit were carried out in this work.

Selected Results

With the MP-PIC methodology, the predicted products yield and temperature distributions are in a very good agreement with the experimental data, with all relative errors below 3.31%. Figure 1a presents the distributions of Cr^{6+}, Cr^{3+}, and Cr^{2+} on catalysts. As it can be seen from Figure 1a, while Cr^{3+} and Cr^{2+} are mainly distributed in the PDH fluidized bed reactor, Cr^{6+} is distributed in the whole CFB setup. Figure 1b shows the flow diagrams of original and updated processes. Numerical results indicate that compared to the original PDH process, the distribution of Cr^{6+} in the whole setup can be inhibited if a pre-reduction reactor is added into the process.

Key Conclusions

Numerical predictions with MP-PIC methodology show a very good agreement with experimental data. According to the full-loop MP-PIC simulations, catalysts covered with Cr^{6+} are distributed in the whole CFB unit. Therefore, effective measures must be taken to avoid chromium contaminates. For instance, a reactor for Cr^{6+} reduction can be set before the PDH reactor in the continuous process.

Acknowledgements

Financial supports from the National Natural Science Foundation of China (21908186) and the Natural Science Foundation of Shandong Province (ZR2023MB109) are acknowledged.

215d Novel Techniques for Surrogate Based Global Sensitivity Analysis on Low Pressure Fluidized Bed Encapsulation

Anurag Guha, Aydin Sunol

This paper presents a novel approach to quantify sensitivity indices for the input parameters used in a low pressure dynamic fluidized bed encapsulation system by surrogate model based global sensitivity analysis (GSA). The physics- based model involves different kind of input variables which can be both spatial and temporal in nature. The input variables are related to fluidization media properties, atomization air and coating solution properties, size and scalability of the bed etc, whereas the output variables are coating thickness of the final product and the coating efficiency. A two-fold scheme can be developed for GSA, where the effect of input parameters can be tested on the dimensionless numbers associated with fluid dynamics, heat and mass transfer, and later those can be mapped further to the final output variables on the basis of coating performance. The physics model-based GSA involves rigorous computational cost due to repetitive model evaluations for sample generation using Monte Carlo simulation, especially when the model has a large number of input parameters. To compensate for that, a data driven approach is required in the form of surrogate modeling to simplify the physics-based model. Three different surrogate modeling techniques have been discussed namely: polynomial chaos expansion (PCE), support vector regression (SVR) and neural network based radial basis function (RBF). The accuracy and efficiency of each of these methods have been compared in the scope of the fluidized bed encapsulation model on the basis of Sobol indices.

215e Analysis of Dry Reforming of Methane Underdifferent Fluidization Regimes Using a Multiphaseparticle-in-Cell Approach

Fahad AlOtaibi, Abdallah S. Berrouk

In the present study, the dry reforming of methane (DRM) has been simulated in fluidized-bed reactors using the multiphase particle-in-cell model. The model was meticulously built to investigate the effect of a wide range of superficial gas velocities covering particulate, aggregative, and lean-phase flow regimes on bed hydrodynamics, conversion, and yields of product gases. Constant values for catalyst loading, CH_4:CO_2:N_2 ratio (1:1:1.3), and catalyst and gas properties were maintained in all simulations. The simulation results obtained are in good agreement with the experimental data reported in the literature. The results show that under different gas velocities, conversion is relatively indiscernible in the particulate regime. In contrast, for the inhomogeneous phases, the turbulent-fluidized bed had the best reactor performance with high CH_4 and CO_2 conversion rates, good CO þ H_2 productivity, and high CO/H_2 molar ratio. This is due to the vigorous turbulent flow and relatively high gas–solid contact. Due to gas bypassing and backmixing triggered by bubbling, the bubbling-fluidized bed generally had the worst performance and below that of the fast-fluidized bed. The present study demonstrates that the performance of DRM reactions in fluidized-bed reactors is strongly related to the hydrodynamics. Moreover, it shows the significance of gas velocity on DRM conversion, yield, and overall reactor performance.

36b Spray-Dried Powder Hand Sanitiser with a Long-Lasting Effect

Viola Tokarova, Lucie Vecerkova, Lucie Maskova, Ondrej Kaspar

Pathogens are spread by contact from person to person, e.g., by handshakes, or indirectly through contaminated surfaces. One way to limit transmission is to keep our hands clean, which is achieved by standard hygiene habits with additional usage of various ethanol-based disinfectants, which became more frequent with the Covid-19 pandemic situation in the last couple of years. However, when access to running water is limited or during long-lasting activities (e.g., surgical procedures, rescue operations), dry powder sanitisers activated by human sweat may find their practical use. Although hand disinfectants and sanitizers have a great immediate effect, they do not provide long-term protection once they dry off the skin. Another aspect is that repeated use of some products (typically alcohol-based) may cause skin irritation or eventually more severe health problems. Therefore, the goal of this work is to create a hand sanitiser with a gradual release of the active ingredient, which would provide protection against microorganisms lasting tens of minutes to hours.

For this purpose, we introduce micron-sized particles, which are retained at the papillary lines of our hands after application with a gradual release of the active component(s). The suitability of chitosan microparticles with incorporated active compound chlorhexidine, and with varying degrees of cross-linking have been investigated. The morphology, particle size distribution, and release rate of the chlorhexidine were compared, and selected particles were further tested on microorganisms (*E. coli* and *S. epidermidis)* for immediate and prolonged effects. Particle deposition, adhesion, and abrasion resistance will be tested on a human finger replica, simulating topographical features and characteristics of the skin.

651b The Roles of Surface Chemistry and Particle Morphology in the Formation and Properties of Silica-Polyelectrolyte Microcapsules

Matthew Lertola, Sapir Lifshiz-Simon, Yeshayahu (Ishi) Talmon, Romain Bordes, Michael Persson, Krister Holmberg, Matthew Helgeson, Bradley F. Chmelka

Particle-stabilized (Pickering) emulsions provide a means to prepare robust, hybrid microcapsules through the adsorption of solids at liquid-liquid interfaces. The co-assembly of silica nanoparticles and polymer species at oil-water interfaces is particularly promising for the development of monodisperse, submicron capsules with adjustable mechanical properties as hosts for water-soluble materials. However, understanding the complicated interactions between multiple components at these interfaces remains challenging, especially at an atomic level.

Here, we establish the atomic-scale interactions between silica surface moieties and polyelectrolyte groups and correlate them with interfacial mechanical properties in a decane-water system. The behavior of anionic silica nanoparticles at oil-water interfaces is fundamentally linked to the surface charge density, which can be modified through surface functionalization with aluminum moieties and further tuned with the pH of the aqueous phase. Two-dimensional $^{29}Si\{^{1}H\}$ NMR analyses enable the unambiguous identification of specific polyelectrolyte and silica surface moieties that bind via strong electrostatic interactions and weaker hydrogen bonding to form nanocomposites that adsorb at oil-water interfaces. Furthermore, these analyses reveal key differences in the populations of silica surface moieties between spherical and elongated silica nanoparticles, resulting in distinct charge densities that are influenced by particle morphology. Cryogenic electron microscopy images show differences in the distributions of spherical and elongated particles on the surfaces of water-filled microcapsules. As shown in rheological analyses of a decane-water system, the interfacial viscosity and stiffness can be controlled by adjusting the relative fraction of these particles in the aqueous phase. By unraveling the coupled relationships between surface chemistry, particle morphology, and macroscopic mechanical properties, these results provide design criteria for the preparation of silica-polyelectrolyte microcapsules with optimized composition and properties for a variety of applications.

651c Tunable Light-Matter Interactions in Complex Perovskite-Coated Particles with Multiscale Chirality

Michael Veksler, Jeffery Raymond, Tao Ma, Nicholas A. Kotov

Chiral nanostructures display chirality at multiple scales – from angstroms to microns and display high chiroptical activity. They encompass a wide range of materials including polymers, nanocarbons, metals, ceramics, and semiconductors. However, the synthesis of nanomaterials of specific chemical nature with a specific set of chiral geometry at specific scale remains a difficult Edissonian process. Among them are perovskites - a large class of highly photoactive materials. The synthesis of perovskite particles with tunable chirality at higher scales is limited by their crystallization patterns and molecular precursors. Here we show that generic achiral perovskites can be deposited on pre-made chiral 'hedgehog' particles (CHIPs) producing optically hybrid materials with spectroscopic bands specific of perovskites and chirality specific of the templates. Left- and right-handed circularly polarized light emission was observed from the hybrid particles and is determined by the geometry of the template CHIPs. The spectral position of chiroptical bands and the analysis of the dissymmetry factors indicate that the circular polarization of perovskite luminescence is dominated by the post-emission scattering by the chiral surfaces in CHIPs. The template pathway is convenient for crafting a wide range of nanostructured chiroptical materials and tunable single-particle emitters.

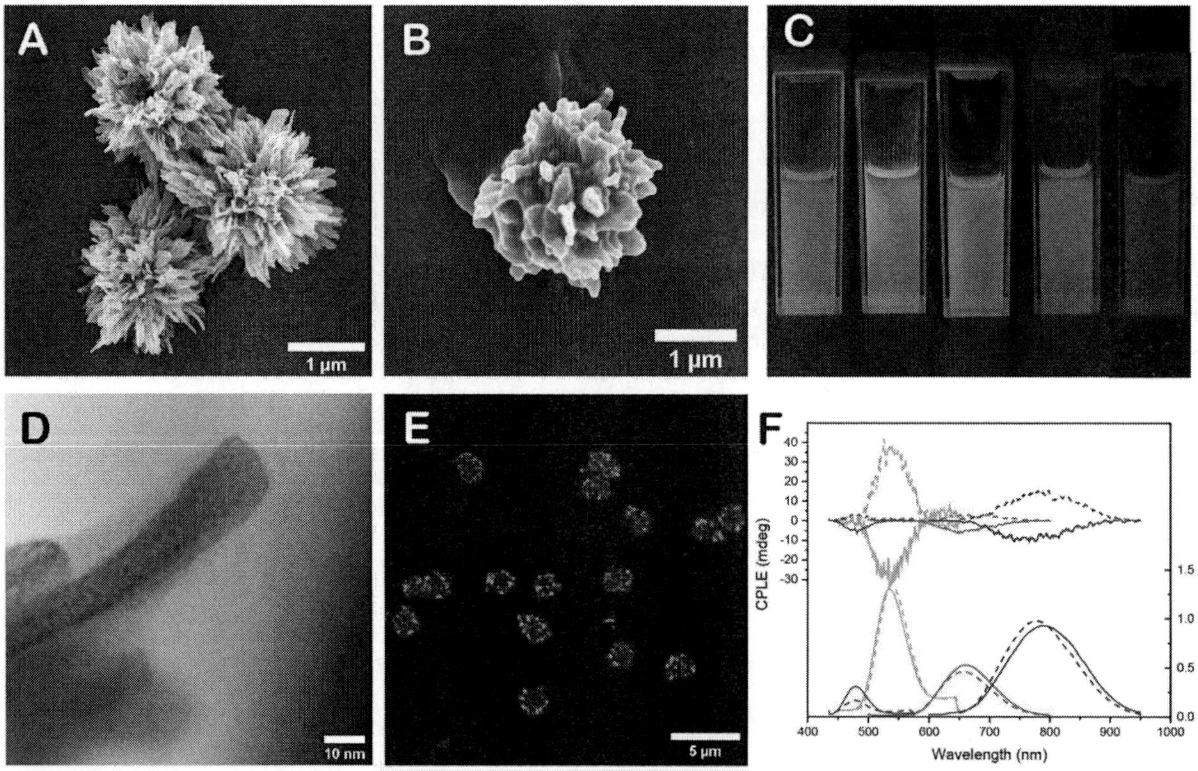

Figure. (**A**) SEM image of CHIPs as-synthesized. (**B**) SEM image of CHIPs coated with luminescent perovskite $FAPbBr_3$. (**C**) Photograph of colloidal dispersion of perovskite-coated CHIPs of varying perovkite composition under UV illumination. (**D**) BF-STEM image of a spike of perovskite-coated CHIPs. (**E**) Z-stack maximum projection of confocal fluorescence micrograph of perovskite-coated CHIPs. (**F**) CPLE of L- (solid) and D- (dashed) CHIPs coated with perovskite.

651d Atomic Layer Etching of Ni-Based Materials By Nitridation

Taylor G. Smith, Ali M. Ali, Jean-François de Marneffe, Jane Chang

Ni and Ni-based alloys are being explored by the semiconductor industry for their use in state-of-the art interconnect technologies and in extreme ultraviolet (EUV) lithography applications. Etching Ni controllably, anisotropically, and selectively at the nanometer scale can be difficult, however, and to date this has limited application of Ni-based materials. Atomic layer etching (ALE) has been used to pattern many metals, but most ALE processes reported to date rely on halogenation or oxidation, which can complicate integration due to corrosion or low etch selectivity.

In this work, an ALE process is presented relying on plasma nitridation, formic acid vapor, and an Ar^+ ion beam. This oxygen- and halogen-free process works by first forming a metastable Ni_xN at the surface using a remote plasma reactor, then removing the Ni_xN layer with the reactive formic acid vapor. The Ar^+ ion beam is used to remove carbonaceous residues left by the formic acid. These three steps constitute an ALE cycle that can then be repeated to etch Ni at a rate of 1.3 nm/cycle. The formation of the Ni_xN as a function of plasma power and pressure is investigated, revealing the role of reactive radicals and ions in the formation of this metastable layer. Selectivity of the ALE process to other materials in semiconductor applications, such as Ru and SiN, is also examined.

651e Autonomous Nanomanufacturing of CsPbBr3 Nanoplatelets

Nikolai Mukhin, Arup Ghorai, Pragyan Jha, Milad Abolhasani

Colloidal metal halide perovskite (MHP) nanocrystals (NCs) have made breakthroughs in a wide range of device applications including displays, light-emitting diodes, and photovoltaics due to their high photoluminescence quantum yield (PLQY) and narrow emission linewidth. The defect tolerance nature of MHP NCs with lower energy bandgaps has resulted in the synthesis of high-performing NCs with high PLQYs (95%+) and narrow emission linewidth [1]. However, as the energy bandgap increases, the surface defects of MHP NCs become more prominent and result in a decrease in their PLQY. In recent years, many reports have attempted to produce high bandgap MHP NCs with the highest PLQY, but have not seen great success compared to lower bandgap MHPs. Conventional synthesis of MHP NCs is performed in batch reactors, but fast formation kinetics can cause irreproducible NCs for both industrial applications and academic research. Continuous flow chemistry adapts to the fast formation kinetics due to the fast, controllable, and reproducible axial and radial mixing occurring in the flow reactors. Along with adapting to the fast formation kinetics, continuous flow chemistry allows for continuous experimentation, in-situ characterization, and reaction miniaturization (lower chemical consumption and waste generation than batch experimentation) [2].

The high-dimensional synthesis space of MHP NCs complicates the process of discovering the best-in-class high-band gap NCs. Self-driving labs (SDLs) with autonomous closed-loop experimentation have gained traction in the past five years for their effectiveness in navigating extremely complex and high-dimensional parameter spaces in a timely manner [3]. Combining both machine learning (ML)-guided experimentation with continuous flow chemistry can decrease chemical consumption and time-to-solution (*i.e.*, identifying the highest-performing MHP NC) by orders of magnitude compared to manual experimentation.

In this work, we present the smart nanomanufacturing of MHP NCs. Specifically, we focus on the autonomous synthesis of high-performing Cesium Lead Bromide ($CsPbBr_3$) nanoplatelets by leveraging an SDL for rapid discovery of their manufacturing route. First, we characterize, validate, and benchmark both the hardware and ML agent of the developed SDL. We then employ autonomous experimentation for the rapid discovery of the synthesis route for high energy bandgap $CsPbBr_3$ nanoplatelets for multiple monolayer thicknesses.

References:

[1] Protesescu, L.; Yakunin, S.; Bodnarchuk, M. I.; Krieg, F.; Caputo, R.; Hendon, C. H.; Yang, R. X.; Walsh, A.; Kovalenko, M. V. Nanocrystals of Cesium Lead Halide Perovskites (CsPbX 3 , X = Cl, Br, and I): Novel Optoelectronic Materials Showing Bright Emission with Wide Color Gamut. *Nano Lett.* **2015**, *15* (6), 3692–3696. https://doi.org/10.1021/nl5048779.

[2] Sadeghi, S.; Bateni, F.; Kim, T.; Yong Son, D.; A. Bennett, J.; Orouji, N.; S. Punati, V.; Stark, C.; D. Cerra, T.; Awad, R.; Delgado-Licona, F.; Xu, J.; Mukhin, N.; Dickerson, H.; G. Reyes, K.; Abolhasani, M. Autonomous Nanomanufacturing of Lead-Free Metal Halide Perovskite Nanocrystals Using a Self-Driving Fluidic Lab. *Nanoscale* **2024**, *16* (2), 580–591. https://doi.org/10.1039/D3NR05034C.

[3] Abolhasani, M.; Kumacheva, E. The Rise of Self-Driving Labs in Chemical and Materials Sciences. *Nat. Synth.* **2023**, *2* (6), 483–492. https://doi.org/10.1038/s44160-022-00231-0.

651f Data-Driven Synthesis Space Mapping of Colloidal Semiconductor Nanocrystals with a Multi-Robotic Platform

Jinge Xu, Fazel Bateni, Chrisropher H. J. Moran, Koray Latif, Andrew Cahn, Milad Abolhasani

All-inorganic Metal halide perovskite (MHP) quantum dots (QDs) have emerged as a highly promising class of semiconducting nanomaterials for various solution-processed photonic devices. These quantum-confined nanocrystals exhibit unique optical properties that can be precisely engineered by altering their composition, shape, size, and geometry(*1*). The surface ligation of MHP QDs relies on an acid-base equilibrium reaction, which is commonly utilized not only to provide colloidal stability in organic solvents but also to tune their optical properties. The use of various organic acids as surface capping ligands results in distinct growth pathways and thereby different QD morphologies(*2*). Consequently, the optical characteristics of MHP QDs are strongly influenced by both the ligand structure (discrete parameter) and the reaction conditions (continuous parameters). The multidimensional nature of this parameter space makes it extremely challenging to comprehensively explore effectively and guide their synthesis. Traditional synthesis methods for MHP QDs, similar to other colloidal QDs, are time-consuming, material-intensive, and laborious, relying on manual flask-based techniques. The manual nature of these methods, along with the interdependent reaction and processing parameters in colloidal QD synthesis, hinders the discovery of optimal formulations and fundamental understanding of MHP QDs(*3*).

In this work, we have developed a multi-robot self-driving lab (SDL) for accelerated synthesis space mapping of room temperature-synthesized colloidal semiconductor nanocrystals. The developed SDL enables systematic automated and autonomous investigation of the effects of ligand structure and precursor concentrations on the photon-conversion efficiency, nanocrystal size uniformity, and bandgaps of MHP QDs. Next, we utilized the developed multi-robot SDL to autonomously map the pareto-front of MHP QDs' optical properties for a rationally selected library of capping ligands.

We overcame challenges of conventional applied and fundamental QD research by investigating the science and engineering of a modular autonomous robotic experimentation platform. We established a closed-loop QD synthesis and development strategy by integrating a modular robotic experimentation platform with data-driven modeling and experiment-selection algorithms. The developed SDL accelerated mapping the optical properties of MHP QDs to the ligand structures and synthesis conditions and understanding the underlying role of ligand structure on the shape, morphology, and optical properties of MHP QDs. The SDL-generated knowledge will enable on-demand synthesis of MHP QDs with optimal optical properties for the next generation energy and display technologies.

Reference:

1. Y. Bai, M. Hao, S. Ding, P. Chen, L. Wang, Surface Chemistry Engineering of Perovskite Quantum Dots: Strategies, Applications, and Perspectives. *Adv Mater* **34**, e2105958 (2022).
2. F. Haydous, J. M. Gardner, U. B. Cappel, The impact of ligands on the synthesis and application of metal halide perovskite nanocrystals. *Journal of Materials Chemistry A* **9**, 23419-23443 (2021).
3. M. Abolhasani, E. Kumacheva, The rise of self-driving labs in chemical and materials sciences. *Nature Synthesis* **2**, 483-492 (2023).

651g Photo-Induced Bandgap Engineering of Metal Halide Perovskite Quantum Dots in Flow

Pragyan Jha, Nikolai Mukhin, Hamed Morshedian, Arup Ghorai, Fernando Delgado Licona, Milad Abolhasani

Over the last decade, colloidal metal halide perovskite (MHP) quantum dots (QDs) have attracted significant attention because of their unique optical and optoelectronic properties, which can be tuned by varying their size, composition, and surface ligands.[1] However, significant challenges persist in precision synthesis and large scale manufacturing of these high-performing ionic nanocrystals. Utilizing a high energy photon flux as the trigger for bandgap engineering of MHP QDs offers a sustainable route for precise bandgap engineering of MHP QDs.[2]

In this work, we present material-efficient microfluidic technology for accelerated *in-situ* studies of the photo-induced bandgap engineering of MHP QDs. Leveraging process intensification enabled by miniaturization, [3] and tunable collimated light source, we rapidly investigate the effect of photon flux and solvent ratio (dihalomethane-to-toluene ratio) on the kinetics and extent of the photo-induced halide exchange reaction of MHP QDs. The developed automated microfluidic technology enables intensification of the photo-induced halide exchange reaction (3.5×) while significantly reducing the material consumption (100×) compared to the conventional batch reactors (*e.g.*, transparent flasks). Additionally, the multimodal *in-situ* characterization probe integrated with the microfluidic reactor enables dynamic investigation of the photo-induced halide exchange reaction without the need for manual sampling. Specifically, we utilized the developed microfluidic technology to study bandgap engineering of cesium lead bromide ($CsPbBr_3$) QDs using dichloromethane as a halogen source ($CsPbBr_{(3-x)}Cl_x$), and unraveled the complex nonlinear mechanism of the photo-induced halide exchange reaction.

The developed microfluidic technology enables, for the first time, detailed mechanistic studies of the photo-induced halide exchange reactions of MHP QDs and opens the door for scalable precision nanomanufacturing of high-performing MHP QDs for applications in displays and photonic devices.

References:

1. Kovalenko, M. V., Protesescu, L. & Bodnarchuk, M. I. Properties and potential optoelectronic applications of lead halide perovskite nanocrystals. *Science* **358**, 745–750 (2017).
2. Wong, Y.-C. *et al.* Color Patterning of Luminescent Perovskites via Light-Mediated Halide Exchange with Haloalkanes. *Advanced Materials* **31**, 1901247 (2019).
3. Morshedian, H. & Abolhasani, M. Accelerated Photostability Studies of Colloidal Quantum Dots. *Solar RRL* **7**, 2201119 (2023).

489a Flame-Made Fractal-like Nanoparticles As Drug Carriers

Georgios A. Sotiriou

Biologics (proteins, peptides, nucleotides) dominate the novel therapeutics market. A major obstacle in their employment, however, is their enzymatic degradation in vivo demanding high doses that result in side-effects [1,2]. One way to avoid degradation of biologics is to encapsulate them in nanocarriers, such as lipid-based nanoparticles. An alternative approach is to use biocompatible inorganic nanoparticles, such as calcium phosphate (CaP). CaP nanoparticles have been explored as drug nanocarriers for biologics and as adjuvants in nanovaccines. In this study, we aim to improve the drug loading capacities of biologics by utilizing inorganic CaP nanoaggregates made by flame aerosol technology, with fractal-like morphology and high specific surface area [3]. We use flame spray pyrolysis, a technique which allows tuning of NPs properties like composition, size, crystallinity. These parameters are critical as they determine the mode of cellular uptake. We optimize the loading of CaP nanoaggreates with biologics, targeting two distinct applications: as antimicrobial peptide delivery carriers and as adjuvants for vaccination. The performance of the developed particles is benchmarked with the state-of-the-art lipid-based nanoparticles for biologic drug delivery.We have synthesized amorphous CaP NPs with varying silica content totune hydrodynamic size, biologic loading capacity, and cell cytotoxicity. We obtain a high specific surface area of greater 150 m2/g for all these NPs. Upon loading ovalbumin, a model protein antigen, we achieved loading capacity values of up to 400 μg/mg NPs. None of these particles showed cytotoxicity on human lung epithelial cells. We further use ovalbumin loaded CaP NPs as a model system to optimize the antigen uptake and tune the inflammatory response by immune cells. The flame-made CaP nanoaggrates developed on this study exhibit advantageous properties for their employment in biologics delivery with high loading values and minimal cytotoxicity.

References

[1] Pati, R., et al. (2018). Frontiers in immunology, 9, 2224.

[2] Lin, Y., et al (2017). Expert review of vaccines, 16(9), 895-906.

[3] Tsikourkitoudi, V., et al. (2020) Molecules, 25(7), 1747.

489b Flame Aerosol Synthesis of Doped Iron Oxide Nanoparticles As Tracers with High Resolution for Magnetic Particle Imaging

Shaquib Rahman Ansari, Andrii Melnyk, Yael del Carmen Suárez López, Carlos Rinaldi-Ramos, Alexandra Teleki

Magnetic particle imaging (MPI) is an emerging diagnostic technique that uses the signal generated solely from superparamagnetic iron oxide nanoparticles (SPIONs). MPI has shown great potential in tumor imaging, cell tracking, and angiography due to its excellent sensitivity, contrast, and safety profile.[1] The biocompatibility of MPI tracers offer additional advantage over other imaging techniques that use harmful ionizing radiation or toxic compounds. However, MPI faces challenges due to its lower resolution compared to other imaging modalities. The MPI signal can be enhanced by tailoring tracer properties such as magnetic moment, saturation magnetization, and anisotropy. An effective approach to modulate these properties is to incorporate elements such as cobalt, nickel, zinc, and magnesium in the SPION lattice.[1] Although, synthesis of MPI tracers with complex stoichiometry presents challenges for large-scale production, it can be circumvented by flame spray pyrolysis (FSP),[2] a scalable manufacturing technique. This study reports a comprehensive investigation of doped SPIONs as MPI tracers while demonstrating the suitability of FSP for their scalable synthesis.

SPIONs doped with manganese and zinc were synthesized by FSP in two crystal size groups (13-16 nm and 22-33 nm) and their MPI performance was evaluated using the MOMENTUM MPI scanner. Post-synthesis functionalization of the nanoparticle surface with citrate and subsequent filtration significantly enhanced the MPI resolution of SPIONs, with Zn-doped nanoparticles exhibiting the highest resolution. These findings signify the influence of hydrodynamic size, size distribution, and SPION composition on the MPI resolution. 2D MPI scans revealed that small-sized Zn-doped SPIONs exhibited optimal MPI performance, characterized by high resolution and sensitivity. The overall MPI performance of all tracers was closely linked to their hydrodynamic size, magnetic diameter, and susceptibility. Finally, as a proof-of-principle, Zn-doped SPIONs were encapsulated in a water-dispersible nanocarrier using flash nanoprecipitation (FNP), circumventing the need for citrate coating while preserving the MPI performance of the tracer. These findings emphasize the significance of doped SPIONs for improved MPI performance, and highlight the combined potential of FSP and FNP in large-scale production of MPI tracers.

Acknowledgement:

This project has received funding from the European Research Council (ERC) under the European Union's Horizon 2020 research and innovation programme (grant agreement No. 101002582). The authors also acknowledge financial support from the Science for Life Laboratory.

References:

(1) Xie, X.; Zhai, J.; Zhou, X.; Guo, Z.; Lo, P.-C.; Zhu, G.; Chan, K. W. Y.; Yang, M.; Xie, X.; Zhai, J.; Zhou, X.; Lo, P.-C.; Yang, M.; Guo, Z.; Zhu, G.; Chan, K. W. Y. Magnetic Particle Imaging: From Tracer Design to Biomedical Applications in Vasculature Abnormality. *Adv. Mater.* **2023**, 2306450.

(2) Mädler, L.; Kammler, H. K.; Mueller, R.; Pratsinis, S. E. Controlled Synthesis of Nanostructured Particles by Flame Spray Pyrolysis. *J. Aerosol Sci.* **2002**, *33* (2), 369–389.

489c Material-Intrinsic NIR-Fluorescence Enables Image-Guided Surgery for Ceramic Fracture Removal

Robert Nißler, Elena Totter, Sebastian Walter, Justus Metternich, Oscar Cipolato, Dimitri Nowack, Alexander Gogos, Inge Herrmann

Hip arthroplasty effectively treats advanced osteoarthritis and has therefore been entitled as 'operation of the 20th century.' With demographic shifts, the USA alone is projected to perform up to 850,000 arthroplasties annually by 2030. Many implants now feature a ceramic head, valued for strength and wear resistance. Nonetheless, a fraction, up to 0.03% may fracture during their lifespan, demanding complex removal procedures. To address this, a radiation-free, fluorescence-based image-guided surgical technique is proposed. The method uses ceramics' inherent fluorescence, demonstrated through chemical and optical analysis of prevalent implant types. Specifically, Biolox delta® implants exhibit strong fluorescence around 700 nm with a 74% photoluminescence quantum yield. We identified emission tails extending into the near-infrared (NIR-I) biological transparency range, forming a vital prerequisite for the label-free visualization of fragments. This ruby-like fluorescence can be attributed to Cr within the zirconia-toughened alumina matrix, enabling the detection of even deep-seated millimeter-sized fragments via camera-assisted techniques. Additionally, fluorescence microscopy detects µm-sized ceramic particles, enabling debris visualization in synovial fluid or histological samples. This label-free optical imaging approach employs readily accessible equipment and can seamlessly transition to clinical settings without significant regulatory barriers, thereby enhancing the safety, efficiency, and minimally invasive nature of fractured ceramic implant removal.

489d A Computational Approach to Predicting the Self-Assembly of Particle-Based Steroid Therapeutics

Oluwaseun Akanbi, Thi Vo, Omolola Eniola-Adefeso

Since their clinical introduction over 70 years ago, steroid-core drugs have been used in several areas of medicine. For example, corticosteroids, such as dexamethasone and prednisolone, are used in the treatment of Covid-19, arthritis, and many other inflammatory conditions. Also belonging to the same steroid-core family of drugs as corticosteroids, bile acids, such as cholic acid and deoxycholic acid, are widely administered to patients with, but not limited to, gall stones, liver diseases, and excessive submental fat.

However, current formulations of these solubilized steroid-core therapeutics suffer from a high clearance rate upon administration, resulting in low bioavailability. Unfortunately, multiple dosing or higher doses lead to local or off-target side effects, for example, ulceration and suppression of the immune system, respectively and noncompliance of patients.

To combat the current challenges of steroid-core therapeutics, particle-based formulations of some bile acids – specifically cholate, deoxycholate, ursodeoxycholate, lithocholate, and chenodeoxycholate – and corticosteroids, namely, methylprednisolone, hydrocortisone, and dexamethasone, was developed using the double solvent emulsion evaporation (DSEE) technique. In this work, the tunability of steroid particles' sizes was demonstrated, which is an important parameter in the administration of particle-based drug delivery systems. Particle tunability was also confirmed with a preliminary coarse-grained computational simulation of particle self-assembly, which somewhat agreed with the experimental results. It was also discovered that despite the modification of deoxycholate particles' sizes, their bioactivity and biodegradability were not negatively impacted.

This study further sought to accurately elucidate the self-assembly mechanism of steroid-core particles and build a library of clinically relevant steroid-core drugs. The systematic coarse-grained model used in this work captures the individual molecular contributions of the steroids' core (base structure) and the extendable tail. Furthermore, the resulting molecular packing of steroid monomers shown in this work agrees with the experimental result to a higher degree than that previously reported. Similarly, this improved systematic coarse-grained modeling was extended to predict the particles' shapes and sizes for other steroid-core drugs if fabricated with the DSEE technique. The library of steroid-core drugs built comprises of Spironolactone, proposed for the treatment of heart diseases, Formestane (Breast cancer), Trodusquemine (Diabetes), Prednimustine (Leukemia), Abiraterone (Prostate cancer), Onapristone (Breast cancer), Aramchol (Fatty liver disease) and Caprospinol (Alzheimer's disease).

Without a doubt, our ability to accurately predict the self-assembly of steroid-core molecules with the coarse-grained computational model offers a cost-effective method of obtaining steroid-based drug particles in real-time. Also, we have previously shown that the sizes of particles fabricated with deoxycholate, a steroid-core molecule, can be altered by modifying its side chain. Therefore, our robust computational model can potentially be useful for determining the suitable side group modification necessary to achieve the desired particles' size, depending on the clinical application. Thus, this study opens opportunities for clinically combating numerous conditions using particle-based steroid therapeutics.

489e Formulation, Efficacy, and Delivery of Organic Nanocarriers Encapsulating Streptomycin for Treating Trees Infected with Citrus Greening Disease

Luiza Oliveira, Kenedy Sanchez, Riley Jones, Hagai Kohay, Philippe Rolshausen, Arnold Schumann, Gregory V. Lowry, Kurt Ristroph

Background: Citrus greening disease, caused by the phloem-limited *Candidatus* Liberibacter asiaticus (*C*Las), is one of the most destructive diseases of citrus worldwide, with currently no effective management strategies or known resistances. Antibiotics, such as streptomycin and oxytetracycline, have been shown to reduce *C*Las titer in infected citrus trees following trunk injection; however, the uptake and subsequent phloem-loading of these antibiotics in a practical, time-efficient, and widely used manner (i.e., via foliar spraying) has been largely unsuccessful with currently available products, even in the presence of uptake adjuvants. In a previous study, organic-coated nanocarriers (NCs) were shown to be phloem-loaded and translocated systemically following foliar administration in tomato plants, motivating the development of surface-similar NCs encapsulating antibiotics for targeted delivery to the phloem of citrus greening-infected trees.

Methods: Encapsulation of streptomycin (STP) into organic NCs with biocompatible surfaces (lecithin, cellulose, and sucrose) was achieved via Flash NanoPrecipitation (FNP) coupled with hydrophobic ion pairing using different counter ions (sodium oleate, dodecyl sulfate, and dodecylbenzosulfonate) and a hydrophobic co-core of vitamin E acetate (Inset 1 in Figure 1). Encapsulation efficiency (EE) and *in vitro* release of STP from the NCs in simulated apoplastic fluid (SAF) over 14 days was quantified via HPLC. The efficacy of the NCs against *L crescens*, a culturable analog of *C*Las, was assessed via optical density (OD) assays (Inset 2 in Figure 1). Lab-scale uptake and translocation (Inset 3 in Figure 1) of surface-similar NCs containing dye or metal in citrus plants was assessed via confocal microscopy and Inductively Coupled Plasma Mass Spectrometry (ICP-MS), respectively. Scale up of NC production was accomplished using syringe pumps and mass flow controllers.

Results and Implications: Streptomycin was encapsulated at high efficiency (EE > 96%) in a variety of nanocarrier formulations with a range of sizes (70 to 250 nm) and STP concentrations (125 to 1000 ppm). Different counterions, charge ratios, and presence of co-core provided controlled *in vitro* release of STP from NCs in SAF varying from 2 to 50% of the theoretical maximum over two weeks. Low release of STP from the NCs is an important first step to demonstrate that STP stays associated with the NC, and hence, would reach the same plant tissues as the NCs do after application. Lab-scale translocation studies done on small (50 cm height) citrus trees have shown that lecithin-coated NCs encapsulating a metal probe were the only ones that reached the root tissues following foliar spraying. Approximately 7% of the NCs detected were found in the roots and 33% in the stem, providing an estimated NC concentration in the phloem of these tissues of 2 to 10 ppm (equivalent to 0.4 to 2 ppm of STP). Additionally, all NC formulations were as efficacious against *L crescens* as free STP (at concentrations ranging from 500 to 0.05 ppm) *in vitro* over 7 days, demonstrating no loss of antibiotic activity due to encapsulation. These preliminary results are a promising indication of the potential of encapsulated STP to be effectively delivered to the phloem of citrus-greening infected trees to combat *C*Las. A year-long field trial on 5-year-old OLL and Hamlin citrus cultivars is underway to test the performance of the best STP NC on actively diseased trees (Inset 4 in Figure 1).

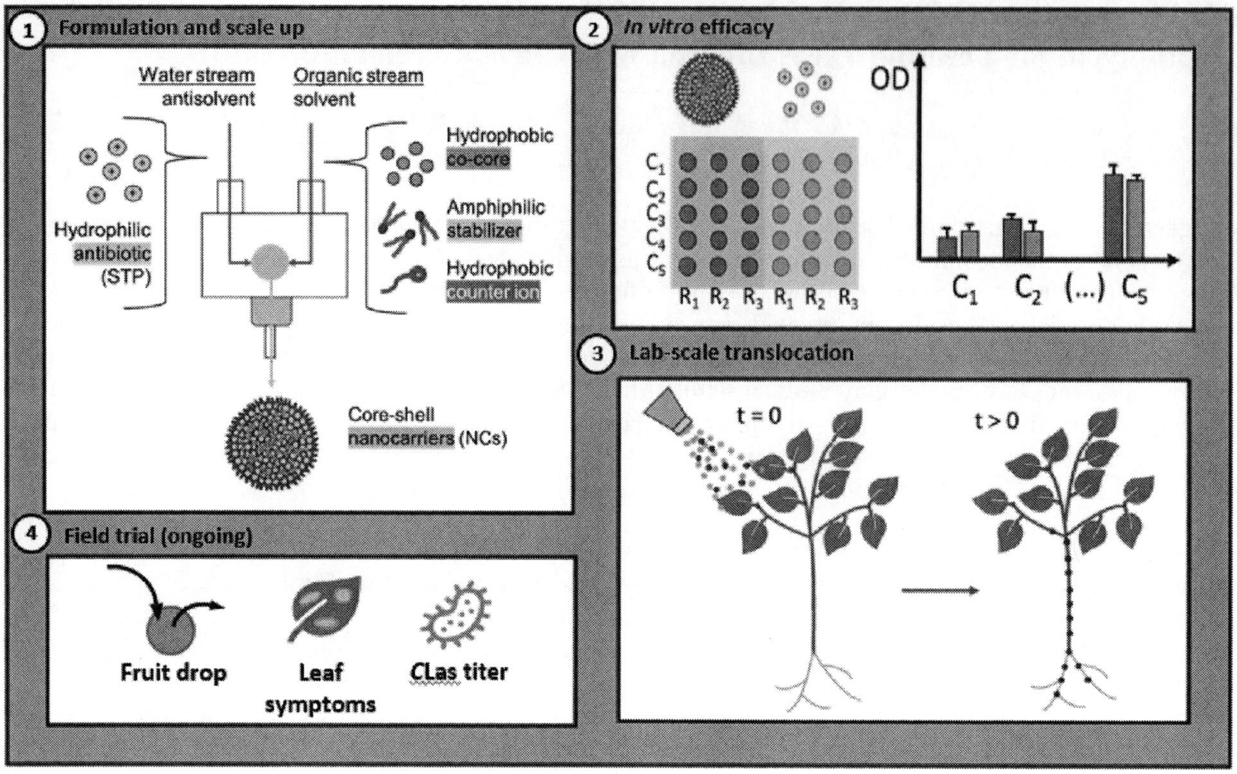

1 Formulation and scale up
Water stream
antisolvent
Organic stream
solvent
Hydrophilic
antibiotic
(STP)
Hydrophobic
co-core
Amphiphilic
stabilizer
Hydrophobic
counter ion
Core-shell
nanocarriers (NCs)
2 In vitro efficacy
OD
C1
C2
C3
C4
C5
R1 R2 R3 R1 R2 R3
C1 C2 (...) C5
3 Lab-scale translocation
t = 0
t > 0
4 Field trial (ongoing)
Fruit drop
Leaf
symptoms
CLas titer

489f Flame-Made Fluorescent Nanothermometers Enable Superior Control in Laser Tissue Soldering

Oscar Cipolato, Inge Herrmann

Laser tissue soldering offers a promising departure from traditional sutures and staples by employing laser light in conjunction with a specialized adhesive material known as solder paste. This innovative surgical technique forms robust watertight bonds, primarily using protein-based adhesives. Its adaptability for minimally invasive procedures presents distinct advantages over conventional methods, which are often challenging to execute in such settings and can lead to complications like inflammation, infection, and fluid leakage, particularly in delicate tissues such as blood vessels, nerves, and the intestine.

However, the efficacy of laser tissue soldering heavily relies on precise temperature control throughout the procedure. Our research introduces a novel approach to this challenge, employing controlled laser tissue soldering guided by fluorescent nanothermometers. These flame-made nanoparticles, seamlessly integrated into protein-based solders, offer non-invasive temperature monitoring. As a significant advancement towards more precise and less invasive soldering techniques, we demonstrate temperature distribution measuring and fluorescence guidance of the soldering process for high-performance outcomes. This is made possible by incorporating an intelligent solder paste containing near-infrared (NIR) fluorescent nanothermometers and nanoabsorbers like titanium nitrate nanoparticles.

Furthermore, our work includes various strategies to guide surgery by analyzing the temperature distribution based on nanothermometry and fluorescence during soldering, thus optimizing results.

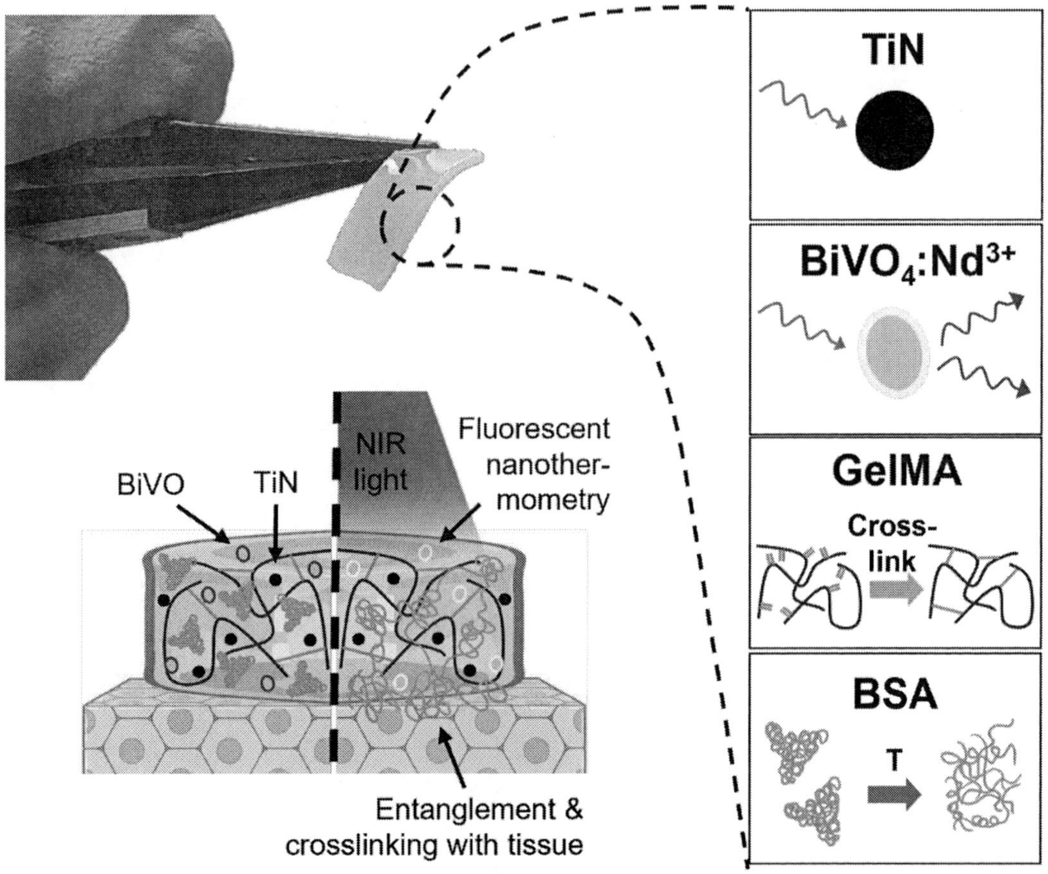

489g Programmable Cargo Release from Jet-Printed Microgel Particles Via an in Situ Ionic Exchange Method

Rong Ma, Sungwan Park, Albert Liu

Hydrogel-based delivery systems have shown promise in clinical settings. They offer precise spatial and temporal regulation of the release of various therapeutic agents, ranging from small molecules and macromolecules to cells. However, achieving programmable release of substances with varying molecular weights at distinct rates remains a challenge. In this study, we develop a scalable method to produce alginate hydrogel particles with finely tuned microstructures capable of controlled cargo release. Our method combines an established jetting technique with a post-synthesis ion-exchange process to finetune the cross-linked alginate microstructure. By varying the type and concentration of ions used, the resulted chemical and physical characteristics of the hydrogel particles can be systematically adjusted, including their swelling behavior and crosslinking density. These modifications enable the alginate hydrogel particles to release multiple model compounds at varying rates. Our findings offer a new strategy for post-synthetic manipulation of the structure and properties of hydrogel microparticles, with many opportunities for biomedical and environmental applications.

86a Materials across Scales for the Energy Transition

Dionisios Vlachos

The energy transition involves complex feedstocks such as biomass and plastics. It requires new processes and materials, efficient ways to deliver heat to reactors, clean hydrogen production, process intensification, and electrification. We will present an overview of some of the challenges of the energy transition and discuss electrification technologies, such as microwaves and Joule heating, toward decarbonizing the chemical industry with an emphasis on high-temperature processes and scalability of the technologies for retrofitting existing infrastructure. We will demonstrate these technologies with examples from shale gas upgrade and plastics recycling.

180n Extended Pt Alloy Ultrathin Surface As ORR Electrocatalysts

Mingze Sun, Zhiqiang Niu

Single crystal surfaces with highly coordinated sites very often hold high specific activities toward oxygen reduction reaction (ORR) and others. However, transferring these high specific activities to practical, high-surface-area electrocatalysts poses significant challenges. In this study, we successfully achieved free-standing extended Pt alloy catalysts by epitaxially depositing Pt alloy on (100)-oriented Cu substrate. The extended Pt alloy surface not only reduces the population of low-coordination sites (e.g., edge and corner atoms prevalent in NPs), but also exhibits 3.1–6.9% compressive strain. The bulk-like characteristics of extended surface is evidenced by site-probe reactions and various spectroscopic analyses. Notably, its ORR activity exceeds that of bulk $Pt_3Ni(100)$ and Pt(111), showcasing a 19-fold increase in specific activity and a 13-fold increase in mass activity relative to commercial Pt/C. Moreover, the electrochemically active surface area (ECSA) is increased by 4-fold compared to traditional thin films (e.g. NSTF), which makes the catalyst more tolerant to voltage loss at high current densities under fuel cell operation. This study broadens the family of extended surface catalysts and highlights the knowledge-driven approach in the development of advanced electrocatalysts.

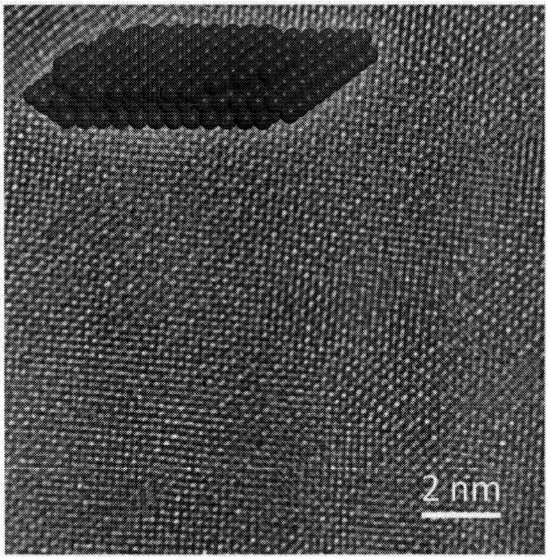

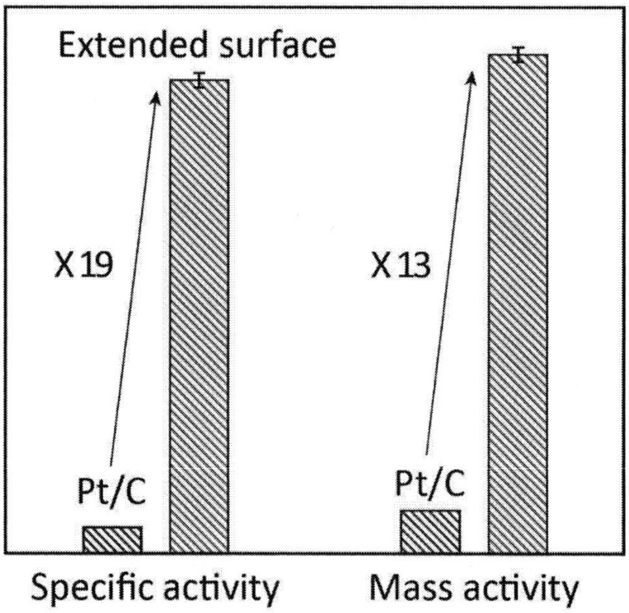

86c Morphology Controlled Synthesis of Catalytic Metal Nanocrystals within 2D Material Nanoconfinements

Joshua Little, Amy Chen, Ali Kamali, Tanmay Akash, Chan-Soo Park, Dongxia Liu, Siddhartha Das, Taylor J. Woehl, Po-Yen Chen

The synthesis of products within confined nanoscale regions, termed nano-confined synthesis, has led to the pioneering of a new generation of heterostructured materials. Within the Van der Waals (VdW) gaps of 2D nanomaterials (2DM), a variety of products can be formed with confined synthesis generating unique sizes, geometries, crystallinities, and even catalytic activities due to the enforced anisotropic growth. However, there remains a gap in knowledge in how to precisely control the complexation of metal ions and strategically guide the in-plane crystallization/growth of metals. Additionally, there is no control over surface chemistry and interlayer spacing of 2DM hosts. Herein, a noble metal ion-complexed assembly of graphene oxide (GO) and Ti3C2Tx (MXene) nanosheets was developed through vacuum filtration. This fabrication method generated nano-confined films with adjustable noble metal loadings. Tetraammineplatinum(II) nitrate (TPtN) was able to serve as an efficient ionic crosslinker to quickly assemble an array of different loaded films. To understand how different water removal processes influence the redistribution of noble metal ions on 2DM substrates, two different solvent drying processes were evaluated in 2DM VdW gaps 1) Air drying at ambient conditions and 2) Freeze drying at 190 K. The TPtN–2DM multilayers were then thermally reduced to synthesize Pt–2DM heterostructures. Then, Pt particle size and morphology were evaluated using HAADF-STEM imaging for all samples. This uncovered that the Pt morphologies were different on each substrate despite the same reduction conditions showing high aspect ratio sheets on MXene versus small cluster/single Pt atoms dispersed across reduced GO (rGO). To understand the observed Pt morphologies, molecular dynamic (MD) modeling was performed using the Large-scale Atomic/Molecular Massively Parallel Simulator (LAMMPS) software to confirm the drying influence on TPtN precursor movement between 2DM sheets by determining equilibrium states of TPtN–2DM systems with and without water. Freeze drying the films result in a greater distribution of noble metal ions within 2DM multilayers, while air drying causes highly concentrated regions of noble metal precursors due to capillary effects of water leaving. Additionally, on MXene surfaces the TPtN ions localize on one side of the confinement while on rGO the TPtN distributes on both sides. Lastly, the Pt–2DM heterostructures showed promise towards the hydrogenation of phenylacetylene. Pt–MXene heterostructures demonstrated a very high substrate dependent activity (2617 molPA molPt-1 h-1) with an average styrene selectivity (82.4%) while Pt–rGO demonstrated a lower activity (1070 molPA molPt-1 h-1) but improved styrene selectivity (85%).

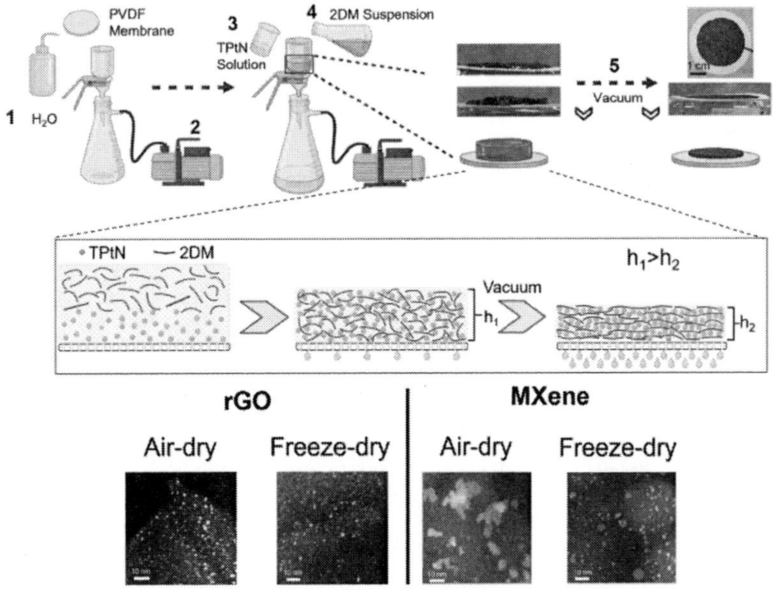

86d Rational Design of Nanoscale Stabilized Oxide Catalysts for the Oxygen Evolution Reaction with OC22

Richard Tran, Liqiang Huang, Yuan Zi, Shengguang Wang, Benjamin M. Comer, Xuqing Wu, Stefaan Raaijman, Nishant Sinha, Sajanikumari Sadasivan, Shibin Thundiyil, Kuldeep B. Mamtani, Ganesh Iyer, Lars Grabow, Ligang Lu, Jiefu Chen

The efficiency of H_2 production via water electrolysis is typically limited by the sluggish oxygen evolution reaction (OER). As such, significant emphasis has been placed upon improving the rate of OER through anode catalyst design. More recently, Open Catalyst 2022 (OC22) has provided a large dataset of density functional theory (DFT) calculations for OER intermediates on the surfaces of oxides.[1] When coupled with state-of-the-art graph neural network models, total energy predictions can be achieved with a mean absolute error as low as 0.22 eV. In this work, we interpolated a database of the total energy predictions for all slabs and OER surface intermediates for 4,119 oxide materials in the original OC22 dataset using pre-trained models from the OC22 framework. This database includes all terminations of all facets up to a maximum Miller index of 1 with adsorption configurations for O* and OH*. To demonstrate the utility of this database, we address one of the major pitfalls in electrocatalyst design: Pourbaix or aqueous stability under harsh reaction conditions. Optimal OER activity generally necessitates reaction conditions such as moderately high temperatures, applied potentials and acidity, all of which disqualifies a large pool of candidate anode catalysts due to aqueous instability. Herein, we propose nanoscale stabilization as a possible solution to expand the pool of existing candidates by improving Pourbaix stability under the prerequisite reaction conditions. To do so, we design a screening framework that can filter materials based on overpotential, surface stability, and stability at the nanoscale using the predicted total energies in our database. From our assessment, we were able to identify 48 and 69 viable candidates for OER under the bulk and nanoscale regime respectively.

[1] Tran et al., *ACS Catal.* 2023, 13, 5, 3066 - 3084

86e CO_2 Hydrogenation to Methanol Using Hybrid Flame-Made CuO/ZrO_2 – Polymer Membrane Reactors

Quang Huy Pham, Eirini Goudeli, Colin Scholes

Methanol exhibits a range of advantages as a potential future fuel due to its high energy density compared to that of natural gas and hydrogen. Traditional methods of methanol synthesis by natural gas have a significant greenhouse gas footprint as they involve CO_2 release as a byproduct. An alternative approach to methanol production by natural gas, is CO_2 hydrogenation. However, high conversion of carbon dioxide into methanol presents significant challenges, primarily arising from thermodynamic constraints and catalyst deactivation due to the formation of water as a by-product. So, research interest has focused on the development of new catalysts, such as Cu/ZrO_2 (Kattel et al., 2016), $Cu/ZnO/ZrO_2$ (Witoon et al., 2018), and $Cu/ZnO/Al_2O_3$ (Tada et al., 2017), that are both active to CO_2 hydrogenation and selective to methanol.

In response to this challenge, hybrid nanoparticle-membrane nanocomposites are manufactured in this work via direct one-step deposition. CuO/ZrO_2 nanoparticles are produced by flame spray pyrolysis (FSP) at various precursor solution-to-oxygen (P/O) flow rate ratios and are deposited on polymeric membranes highly selective to methanol (Figure 1a). Figure 1b and c shows scanning electron microscopy (SEM) images of flame-made CuO/ZrO_2 catalyst produced at a P/D ratio of (b) 2 mL/min precursor solution to 8 L/min oxygen (2/8, cold flame) and (c) 10/8 (hot flame), deposited on 6FDA-TMPDA membranes. Increasing the precursor flow rate, leads to higher temperature and longer high-temperature particle residence time, which in turn results in larger CuO and ZrO_2 crystal size and smaller specific surface area.

The effect of nanoparticle and film characteristics on the CO_2 conversion and methanol production rate is quantified. FSP conditions leading to nanocatalysts with highly activity to CO_2 hydrogenation are identified, while the selectivity to methanol, which is enhanced by the presence of polyimide membranes, is also evaluated. At 200 °C and 20 bar, these catalytic membrane reactors demonstrate exceptional performance, achieving 113% improvement in carbon dioxide conversion and a remarkable 106% increase in methanol production rate (Figure 1d) compared to conventional catalytic fixed bed reactors (Fig. 1d: horizontal line). Overall, the catalyst-membrane reactors have shown long-term stability and ability to suppress catalyst deactivation caused by water, ensuring sustained catalyst stability throughout the reaction.

Such hybrid membrane-catalytic nanoparticle reactors are a promising technology which enables industries to convert their carbon dioxide wastes into valuable energy products, such as methanol.

Funded by Future Fuels Cooperative Research Centre grant (RP1.3-04) and the University of Melbourne, Australia.

References

Kattel, S., Yan, B., Yang, Y., Chen, J. G., and Liu, P. (2016). *J. Am. Chem. Soc.* 138: 12440-12450.

Witoon, T., Numpilai, T., Phongamwong, T., Donphai, W., Boonyuen, C., Warakulwit, C., Chareonpanich, M., and Limtrakul, J. (2018). *J. Chem. Eng.* 334: 1781-1791.

Tada, S., Watanabe, F., Kiyota, K., Shimoda, N., Hayashi, R., Takahashi, M., Nariyuki, A., Igarashi A., Satokawa, S. (2017). *J. Catal.* 351: 107-118.

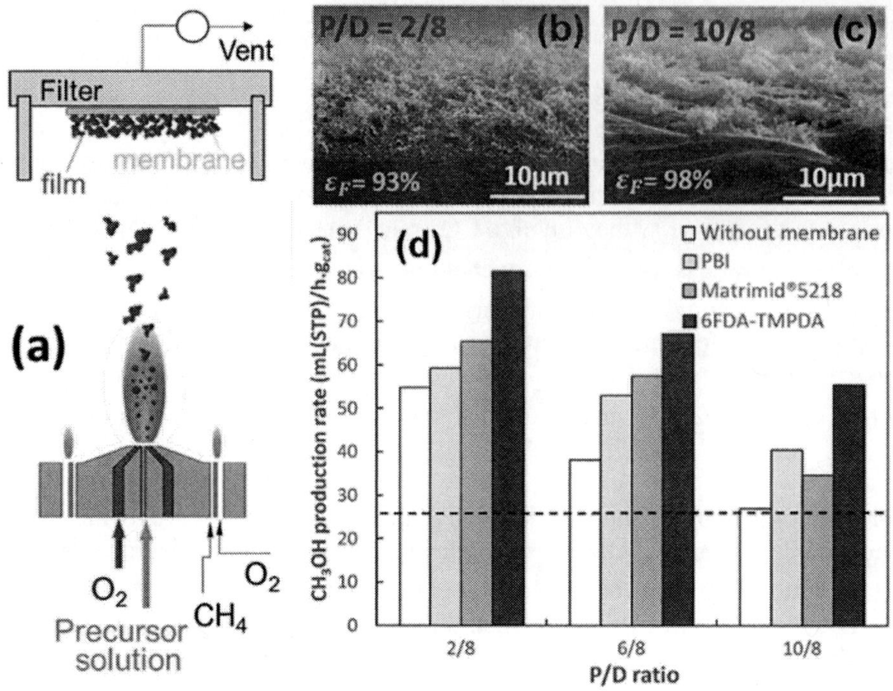

Figure 1. (a) Experimental setup for direct, one-step deposition of flame-made nanoparticles on polymeric membranes. SEM images of CuO/ZrO_2 nanoparticles produced in (b) 10/8 and (c) 2/ flames, deposited on 6FDA-TMPDA. (d) Effect of flame process conditions and polymer substrate on methanol production rate.

86f Tuning Metal Nanoparticle Surface Site Accessibility, Electronic State, and Reaction Microenvironment Using Bound Organic Ligands

Sayed Abu Sufyan, Michael Nigra

Synthesis of metallic nanoparticles in solution approaching 1 nm in diameter which contain accessible surface sites with covalently bound organic ligands remains a challenge. We demonstrate the synthesis of a series of gold nanoparticles, ≤ 2 nm in diameter, using bulky bisphosphine ligands, which impart stability to the nanoparticles. Au nanoparticles stabilized by bis(diphenylphosphino)methane (dppm) exhibited the highest accessibility, where 61% of the total gold atoms bind 2-napthalenethiol and this represents the highest fraction of accessible atoms reported for an organic ligand-stabilized metallic nanoparticle in solution. The Au nanoparticles synthesized with bisphosphine ligands are all active for resazurin reduction reaction, and the reaction rate scales with the number of binding sites on the nanoparticle surface.

There is additional potential in modulating the electronic characteristics of the metallic nanoparticles through bound organic ligands with different functional groups, such as electron-donating phosphines and electron-withdrawing thiols. We synthesize and compare ligand-bound Au nanoparticles of the same size with structurally similar ligands but with thiols (triphenylmethyl mercaptan, TPMT) and phosphines (triphenylphosphine, TPP) to assess the electronic effects of the ligand. The similarly sized nanoparticles showed a similar number of accessible sites and exhibit comparable catalytic activities for benzyl alcohol oxidation and CO oxidation. However, when the TPP- and TPMT-stabilized Au nanoparticles are used in resazurin reduction, the TPP-stabilized nanoparticles are active, while the TPMT-stabilized nanoparticles are inactive. It is hypothesized that this on/off behavior in activity, which is not due to differences in nanoparticle size or number of active sites, originates in electronic differences of the Au nanoparticle surfaces as supported by evidence from X-ray photoelectron spectroscopy and FTIR of bound CO molecules.

To further investigate the role of these ligands in modulating the electronic properties and reaction microenvironment of metal nanoparticles, we synthesize ruthenium nanoparticles using phosphine (both TPP and dppm) and thiol (TPMT) ligands. The average diameter of the Ru nanoparticles is approximately 0.8 nm and these are believed to be the smallest ligand-stabilized Ru nanoparticles synthesized in solution in the literature. The similarly sized Ru nanoparticles showed a similar catalytic activity for CO oxidation. However, in H_2O_2 decomposition, the identity of the ligand (monodentate *vs.*bidentate) does play an important role in this reaction. It is observed that the hydrophobicity/hydrophilicity of the ligand has a significant impact on the H_2O_2 decomposition rate, with a more hydrophobic ligand layer leading to reduced H_2O_2 decomposition rates. Thus, this contribution demonstrates the potential of utilizing bound organic ligands to control H_2O_2 decomposition rates and other reactions which are sensitive to hydrophobic/hydrophilic environments.

86g Understanding the Reaction Mechanism of One-Dimensional Silver and Palladium Nanostructure Synthesis in a Millifluidic Reactor: A Kinetic Study Using UV-Vis Spectroscopy

Shohreh Hemmati, Sina Kaabipour, Chamath Vindula Basnayake Pussepitiyalage, Destiny Williams

The synthesis of one-dimensional (1D) silver and palladium nanostructures holds great promise for various applications due to their unique properties. In this study, we investigate the reaction mechanism involved in the synthesis of these nanostructures using kinetic studies coupled with UV-vis spectroscopy. Our approach involves the polyol synthesis of silver nanowires (AgNWs), green AgNWs synthesis using tannic acid, green palladium nanorods (PdNRs) synthesis using L-ascorbic acid, and PdNRs synthesis via biotemplating of Tobacco Mosaic Virus (TMV), Barley Stripe Mosaic Virus (BSMV), and their Virus-Like Particles (VLPs) in a millifluidic reactor.

By systematically varying reaction parameters such as temperature, reaction time, and precursor types and concentrations, we aim to elucidate the underlying kinetics governing the formation of these 1D nanostructures while maximizing their yields for large-scale industrial relevant production. UV-vis spectroscopy is employed to monitor the evolution of the reaction and composition of the metal precursors in real-time, providing valuable insights into the nucleation and growth processes. Such insights are crucial for the rational design and optimization of these nanostructures for a wide range of applications including catalysis, sensing, and nanoelectronics manufacturing, to name a few.

651a Flame-Made Nanoparticles for Molecular Sensing

Andreas Guentner

Combustion aerosol technology has distinct advantages for the assembly of molecular sensors (e.g. chemoresistive or plasmonic) compared to their traditional wet chemistry synthesis. These advantages are traced to combustion's steep temperature gradients and high particle concentrations during sensing particle formation and film deposition. This gives direct access to a plethora of material compositions (e.g. metastable phases, solid solutions, mixed oxides) and fractal-like porous but rigid structures that can lead to unique sensor selectivity, sensitivity and stability along with short response and recovery times.[1] Here, I will discuss our recent advances on the nanoscale engineering of flame-made molecular sensing materials and filters. For instance, finely dispersed CuO_x clusters on Co_3O_4 will be presented that enable the selective detection of formaldehyde at low operational temperatures.[2] This will be followed by the design of highly reactive noble-metal clusters for plasmonic gas sensing that enable selective air pollutant detection with wearables even by naked eye.[3] I will show how metastable crystal phases are captured that enable unprecedented humidity robustness and selective benzene sensing, even over chemically similar molecules. Finally, devices for food quality assessment[4]and safe green fuels handling will be discussed that have been commercialized recently and sold to customers in 26 countries on 5 continents[5], demonstrating the immediate practical impact of such flame-made sensors.

[1] Güntner, A. T., Pineau, N. J. & Pratsinis, S. E. *Prog. Energy Combust. Sci.* **90**, 100992, doi:https://doi.org/10.1016/j.pecs.2022.100992 (2022).
[2] D'Andria, M., Krumeich, F., Yao, Z., Wang, F. R. & Güntner, A. T. *Adv. Sci.*, 2308224, doi:https://doi.org/10.1002/advs.202308224 (2024).
[3] Güntner, A. T. & Schenk, F. M. *Nanoscale* **15**, 3967-3977, doi:10.1039/D2NR06599A (2023).
[4] Abegg, S., Magro, L., van den Broek, J., Pratsinis, S. E. & Güntner, A. T. *Nature Food* **1**, 351-354, doi:https://doi.org/10.1038/s43016-020-0095-9 (2020).
[5] Güntner, A. T., D'Andria, M. & van den Broek, J. *Nature Reviews Bioengineering* **1**, 385-387, doi:10.1038/s44222-023-00068-y (2023).

36c Design of Modular 3D-Printed Milli-Fluidic Mixers to Enable Sequential Nanoprecipitation (SNaP) for the Tunable Synthesis of Nano- and Micro- Particles

Nouha El Amri, Thomas Belinky, Parker Lewis, Rachel Pollard, Allie Karakosta, Nathalie M. Pinkerton

Introduction:

Sequential NanoPrecipitation (SNaP) is an emerging synthesis process for the controlled and robust formation of nanoparticles for drug delivery and bioimaging. SNaP relies on the same rapid mixing and controlled self-assembly principles as traditional one-step Flash NanoPrecipitation (FNP)[1-3]. However, it involves a two-step assembly process by decoupling the core formation and stabilization[4]. This provides a novel strategy to better control the final nanoparticle structure.

Given the novelty of this synthesis process, the understanding of the nanoparticle formation mechanism remains incomplete, and the impact of the synthesis parameters on the synthesis outcome is yet to be investigated. In particular, the impact of the delay time separating the core formation and the stabilization is lacking. This is due to technical challenges stemming from a lack of versatility in the available SNaP experimental setups. Indeed, current SNaP experimental mixing set-ups rely on the use of two commercially available mixers connected via tubing in a series arrangement[1]. This setup suffers from several drawbacks including significant dead volumes, lack of modularity, and limited mixing geometries. **Our goal** is to present an approach for rapidly prototyping the SNaP mixers to address the issues currently observed with existing setups. These mixers will be used to study the impact of the synthesis parameters on the synthesis outcome. Specifically, we will investigate the impact of the delay time on the nanoparticle's size.

Method:

In this work, we first presented a 3D-printed mixing platform, that relies on the same rapid mixing principle as the currently used rapid mixers, while offering greater modularity and better control over the synthesis parameter. We later used this platform for the synthesis of drug and fluorophore-loaded polymeric particles. We additionally used this platform to study the impact of the synthesis parameters on the synthesis outcome.

We used Dynamic Light Scattering (DLS) for size measurments, High Performance Liquid Chromatography (HPLC) for measuring the drug loading, and ThermoGravimetric Analyzer (TGA) for determining the total particle solids concentration.

Results:

With our 3D approach for the fabrication of the SNaP mixers, we rapidly prototyped several mixers with diverse designs. We first designed two mixing configurations that differ in the number of the inlets of the first mixer stage. We then changed the distance that separates the first mixing stage and the second mixing stage. Such change resulted in a change in the delay time between the core formation and the stabilizer addition. We later demonstrated that this setup enabled us to synthesize drug-loaded nanoparticles through SNaP with high encapsulation efficiency. We then showed that by changing the mixer configuration we were able to access the micron size range, previously inaccessible with SNaP and traditional FNP. We were able to form drug-loaded microparticles with a size of up to 1.2 μm. Additionally, we investigated the impact of the delay time on the synthesis outcome, and we evidenced that this parameter is crucial for controlling nanoparticle size. We observed that by changing the delay time from 5.8 ms to 10.5 the size of the formed nanoparticle increased from 198 nm to 233 nm. This allowed us to expand the size range of the polymeric particles formed through rapid nanoprecipitation.

Conclusion:

This approach presents a strategy to tune the nanoparticle's size while keeping the composition constant. This was achieved with the use of our 3D-printed SNaP mixers that provided control over the various synthesis parameters, and most importantly allowed us to tune the delay time with precise control. This also presents a strategy to have a better understanding of the particle formation mechanism, which can present a strategy to enhance drug loading efficiency, and stability and to tackle certain challenges with current methods used to produce drug-loaded nanoparticles.

1. B. K. Johnson and R. K. Prud'homme, *Australian Journal of Chemistry*, 2003, **56**, 1021-1024.
2. K. M. Pustulka, A. R. Wohl, H. S. Lee, A. R. Michel, J. Han, T. R. Hoye, A. V. McCormick, J. Panyam and C. W. Macosko, *Mol Pharm*, 2013, **10**, 4367-4377.
3. W. S. Saad and R. K. Prud'homme, *Nano Today*, 2016, **11**, 212-227.
4. N. J. Caggiano, S. K. Nayagam, L. Z. Wang, B. K. Wilson, P. Lewis, S. Jahangir, R. D. Priestley, R. K. Prud'homme and K. D. Ristroph, *Int J Pharm*, 2023, **640**, 122985.

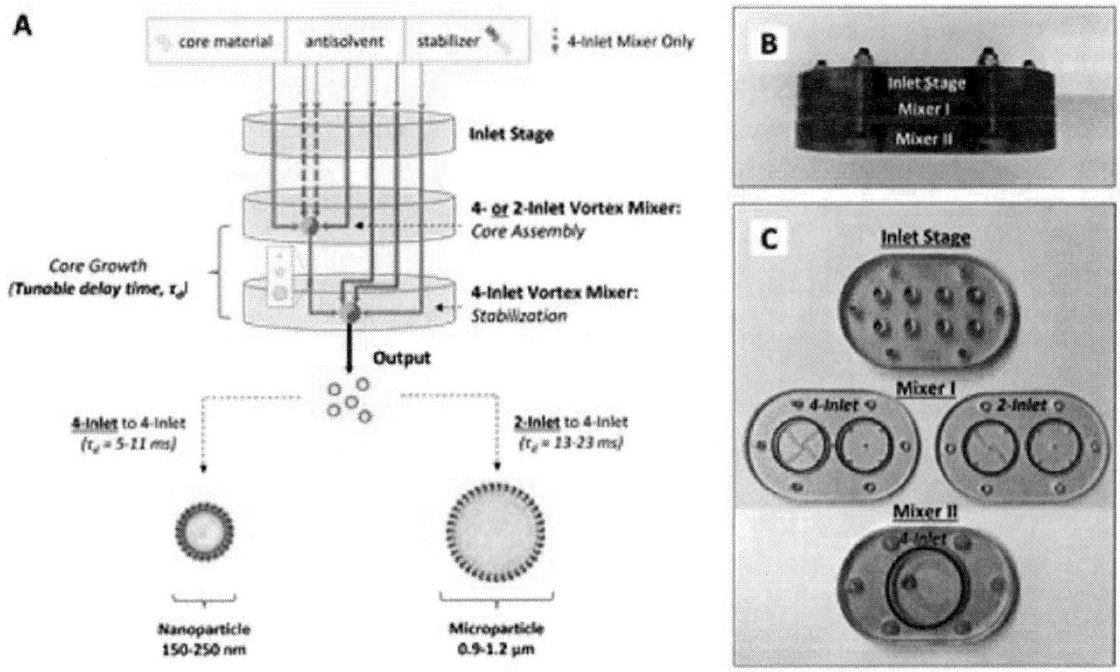

36d Investigating the Inflammatory Response to Exposure of Ultrafine TiO_2 Particulate Matter to Huvecs

Laura A.E. Brunmaier, Travis W. Walker

Epidemiological studies have indicated that strong causal evidence exists to link the inhalation of particulate matter to the exacerbation of pathology in the cardiovascular system, ranging from myocardial infarction and atherosclerosis to direct cytotoxicity and inflammation. Ultrafine particles are ubiquitous in ambient air, in industrial sites, and in air pollution. When particles are inhaled, deposition can occur in the lungs, and the mechanisms of pathology have been well studied. However, ultrafine particulate matter can translocate from the lungs into the bloodstream to circulate throughout the body. Contradictory evidence exists of inflammation and cytotoxicity that is caused from nanoparticle exposure to the endothelium. When endothelial cells (ECs) are adversely stimulated, they have been shown to secrete cytokines that mediate an inflammatory response. Currently, studies that quantitatively evaluated the secretion of pro-inflammatory cytokines from endothelial cells upon nanoparticle exposure are not accounting for aggregation that occurs between particles over time and, therefore, are exposing cells to a wider range of agglomerated sizes. This study evaluates the inflammatory response from ECs after particle exposure. The ECs are then exposed to the particles in a transwell system, where we take advantage of the effects of gravitational settling to expose the ECs only to the smallest fraction of the particles that are in suspension. After72 hours in the transwell assay, we found that the inflammatory response between varying concentrations of particles mirrored the inflammatory response of the positive control of lipopolysaccharide (LPS). These results indicate that the inflammatory response may have a stronger relationship to the particle size than to the concentration of the particles in mass per volume.

36e Sequential Nanoprecipitation (SNaP) Allows Uniform Co-Encapsulation of Small Molecules and Colloidal Nanocrystals within Polymeric Nanospheres

Rachel Pollard, Nouha El Amri, Parker Lewis, Ashley Han, Thomas Belinky, Jacques Zacharie Thaddeus Ponce, Nathalie M. Pinkerton

Flash Nanoprecipitation (FNP) is used in drug delivery for the tunable, scalable, single-step synthesis of polymeric core-shell nanoparticles (NP) encapsulating small molecule active pharmaceutical ingredients (APIs). Recent interest in colloidal nanocrystals as APIs, or in co-administration with APIs, has yielded a high volume of research aimed at synthesizing hybrid, multifunctional nanoparticles that can be used as diagnostic bioimaging probes and deliver therapeutic payloads. Nanocrystals such as quantum dots, gold NP, and magnetic NP have distinct optical, electronic, and paramagnetic properties based on their size, motivating the need to be able to robustly synthesize composite NP that coencapsulate many sizes of nanocrystals and small molecules. While FNP has been previously used to encapsulate hydrophobic colloidal gold and magnetic nanocrystals, the process is limited by the size of the nanocrystals and polymer co-excipient. Additionally, characterization and workup of these composite NP reveal nonuniform loading of nanocrystals and small molecules, and a low "yield" of NP that contain significant loadings of nanocrystals.

Smoluchowski growth kinetics, which have been predicted as governing principles of FNP, predict orders of a magnitude differences in assembly timescales for dense, low diffusivity crystals versus small molecules. Thus, in this work, we aim to circumvent the problem of mismatch by delaying the co-excipient growth and stabilization phases to overcome different assembly timescales.

Herein we utilize 3D printed prototypical multistage mixers to synthesize nanocrystal-loaded polymeric nanospheres using sequential nanoprecipitation (SNaP). By separating the core components into separate inlet stages and tuning resident lengths between stages, we uncover dominating parameters of diffusion-limited growth kinetics. We demonstrate ability to synthesize uniform composite PLA-based nanospheres 200nm in diameter encapsulating quantum dots, iron oxide, and gold nanoparticles from 2nm to 20nm in size. Ultimately, we show 2+ fold increases in nanocrystal loading yields compared to FNP, and we show nanosphere formation can be decoupled from nanocrystal size.

36f Selectivity of Bacterial Capture By Polydopamine-Coated Magnetic Nanoclustsers

William Pitt, Bowen J. Houser, Alyson N. Camacho, Camille A. Bryner, Masa Ziegler, Tochukwu P. Okonkwo, Roger G. Harrison, Karine Chesnel

Fast and precise identification of bacteria is essential for medical diagnostics, and for food and water quality control. Magnetic particles covered with specific surface chemistry that binds to bacteria are efficient at removing bacteria from a sample for further analysis. Herein we present a novel approach for bacterial concentration using nanoclusters of magnetic nanoparticles coated with polydopamine (pDA), a polymer known to bind aquatic mussels to surfaces, but heretofore assumed to be anti-biofouling and perhaps antimicrobial when coated on medical devices. We have discovered that 5-nm thick coating of pDA on magnetic nanoclusters (MNCs) rapidly and strongly binds to many types of bacteria, including those associated with medical infections and food and water contamination. Furthermore, the bacteria remain alive and can grow in the presence of the MNCs. We present our study of binding of several gram-positive and gram-negative bacteria to pDA-coated MNCs, including *Staphylococcus epidermidis*, *Staphylococcus aureus*, *Streptococcus mutans*, *E. coli*, and other common infectious species. These functionalized MNCs are selective in binding various strains of bacteria. For example the binding efficiency to bacteria in a suspension ranges from 0.01 for an *E. coli* strain to 0.98 for *S. aureus*. The fraction of bacteria captured reaches a plateau that is different for various species, indicating that there are specific interactions between the pDA coating and various types of bacteria. This binding and recovery technique has immediate application in rapid capture and subsequent identification of bacteria in infected body fluids (blood, urine, cerebral spinal fluid), hospital environments, food quality assurance, and water purification.

36g Study of Molecular Interaction and Texture Characteristics of Hydrocolloid-Mixed Alginate Microspheres: As a Shell to Encapsulate Multiphase Oil Cores

Yuting Wu, Boya Lv, Shiteng Wang, Zhe Liu, Xiao Dong Chen, Yi Cheng

This work investigates the molecular interaction of hydrocolloids (xanthan gum (XG), hydroxyethyl cellulose (HEC), carbomer (CBM) and hymagic™-4D (HA)) with sodium alginate (SA) in microspheres in detail. The molecular interaction of hydrocolloids with SA are demonstrated by the rheological property analysis of the mixed solutions as well as the morphology structure and texture characteristics studies of the microspheres. It is found that the hydrocolloids (XG, HEC and CBM) with branches or capable to coil are able to form complex networks with SA through molecular interactions which hinders the free diffusion of calcium ions and changes the texture characteristics of microspheres. In addition, the mixed solutions (SA-XG and SA-HEC) with complex networks and do not have a chelating effect on calcium ions are used to form the shell of the microcapsules through droplet microfluidic technology, and stable with soft microcapsules encapsulating multiphase oil cores have been successfully prepared. At the same time, micro texture analyzer (Micro-TA) is introduced to quantify the textural properties of microspheres and microcapsules in relation to human sensory properties. The developed stable and soft microcapsules which have the properties of sensory comfort are expected to be applied in the personal care industry and a variety of fields.

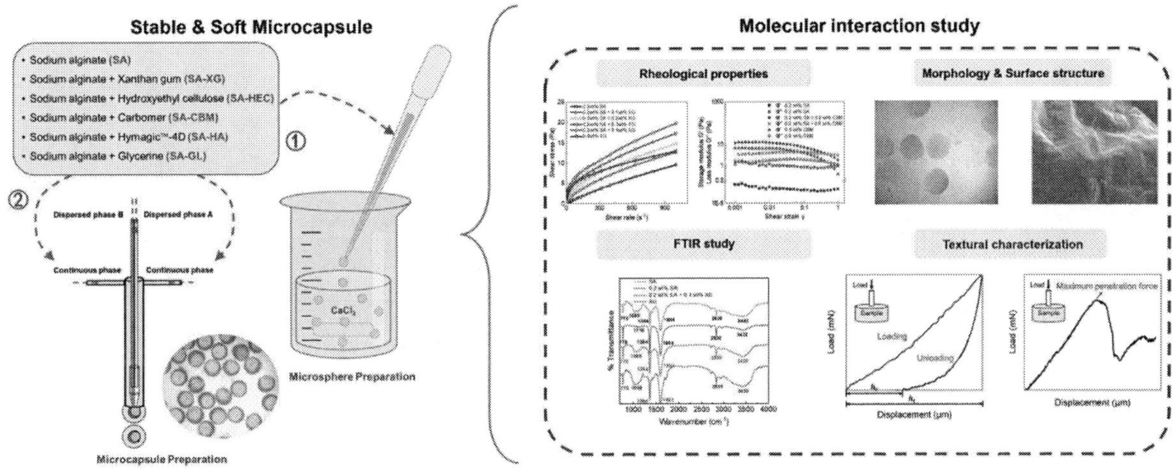

36h Albumin Nanovectors with Tunable Size and Hydrophilicity As a Precursor to Bifunctional Colloids

Blake Bartlett, Sepideh Razavi, John Klier

The success of albumin-bound chemotherapeutics has demonstrated the protein's value as a nonimmunogenic drug platform, significantly boosting residence time compared to naked drugs. One emerging area of research is in bispecific recognition, wherein a single drug particle can bind cancer cells to immune cells and trigger the beginning of an acquired immune response. However, many of these novel treatments suffer from poor bioavailability and low stability, and the necessity of a nanocarrier that maintains the bifunctional morphology is clear. To achieve directionally constrained bifunctional behavior, this research seeks to synthesize 'bilobal' particles where a lipophilic sphere is self-assembled onto a previously synthesized hydrophilic sphere. While synthetic bilobal colloids are well known, there are no known examples of biocompatible bilobal colloid-based proteins such as albumin suitable for in-vivo application. We are synthesizing albumin particles with anisotropy in both shape and surface chemistry allowing for multidentate functionalization of "lobes" independently of one another. Progress to date has optimized synthesis of hydrophilic albumin nanospheres using green chemistry and multiple crosslinkers under ambient conditions, as well as ligand functionalization that alters the hydrophilicity of albumin and provides different binding sites for later conjugations. The aging studies conducted on the synthesized particles illustrated that they are colloidally stable even in biologically isotonic electrolytic conditions for over six months. Manipulation of electrolyte concentrations and ligand concentrations allow for independent control of size, polydispersity, and zeta potential.

216a Some Fundamental Insights from Late Professor Leonard Austin's Research on the PBM of Milling Processes

Ecevit Bilgili

Professor Leonard Austin significantly contributed to the development of grinding–milling unit operation and its modeling in a similar manner to that of the development of other chemical engineering unit operations such as distillation, absorption, etc. Until his retirement in 1996, he was a professor in Fuel and Mineral Engineering at the Pennsylvania State University; he remained an active researcher–engineer until he passed away on Nov. 21, 2022. Interestingly, despite being a mineral engineering professor, his work in population balance modeling (PBM) as well as process development and scale-up of milling processes reflect the style of a first-class chemical engineer incorporating all fundamental approaches and principles of chemical engineering especially those that pertain to residence time distributions (RTDs) and reactor design–kinetics. In fact, his chemical engineering style of process analysis can be easily seen in his most celebrated review paper [1] and his book on ball milling [2], which became a classic in the field. Perhaps as a secondary point to the above, Prof. Austin was a visiting professor in chemical engineering departments, published in chemical engineering journals, and attended AIChE meetings. Hence, it is of no surprise that a session at the AIChE Annual Meeting honors his contributions to the development of comminution processes. It may not be an exaggeration to state that Prof. Austin was one of the most influential figures in the field of comminution processes. Any chemical engineer who modeled comminution processes used the PBM equations in the formalism of Prof. Austin, used his breakage distribution function, and/or used his direct–indirect methods for determining the PBM parameters. Obviously, this presentation cannot do justice to all of Prof. Austin's monumental contributions; therefore, it is not intended to serve as a review. Instead, my main goal is to distill from his work the most important theoretical developments within the PBM framework and discuss the fundamental insights gained that are significant and relevant to this day.

First part of my talk is devoted to Prof. Austin's most celebrated review paper on the treatment of grinding as a rate process [1], wherein wide variety of models developed until 1970 were shown to be inter-convertible versions of a single model, but with differences in formalism. This systematization of the modeling approaches for milling processes was and is of great importance because even to this day chemical, civil, mineral, and mechanical engineers appear to use PBMs for milling processes that mainly differ in their formalism, which makes communication and knowledge transfer between these disciplines less efficient. In other words, Austin's review was the first attempt to unify various PBMs with differing formalisms. In essence, the PBMs for batch/plug-flow milling processes were categorized based on (i) the particle distribution (PSD): cumulative vs. density–frequency distributions, (ii) size-continuous vs. size-discrete forms, and (iii) time-continuous vs. time-discrete forms. Moreover, the size-discrete form of the PBM was presented in both the index notation and the matrix notation.

The second part of my presentation is dedicated to the estimation of the PBM parameters. By and large, all PBMs for batch/plug-flow continuous milling are based on two fundamental functions (kernels) [3,4]: a specific breakage rate parameter S_i and a breakage distribution parameter b_{ij} (or its cumulative counterpart B_{ij}). For a well-mixed batch ball mill or a plug-flow continuous mill at the steady state, the time (retention time)-continuous, size-discrete form of the PBM assuming first-order (linear) breakage is given as follows [5]:

$dM_i / dt = -S_i M_i + \sum b_{ij} S_j M_j$ with $M_i(0) = M_{i,\mathrm{ini}}$ (1)

In Eq. (1), i and j are size class indices that extend from size class 1, which contains the coarsest particles of size x_1, to size class N, which contains the finest particles of size x_N. t is milling time, and M_i is mass fraction in size class i. $M_{i,\text{ini}}$ is the mass fraction of feed material in size class i. To estimate S_i and B_{ij} parameters, Prof. Austin developed various *direct calculation methods* including short duration, single-fraction milling tests and the BI, BII, and BIII methods with increasing accuracy [6,7]. In addition, he spearheaded the optimization-based back-calculation (indirect methods) of the parameters [8,9]. Pros/cons of these different approaches will be discussed in this talk.

The third part of my presentation is dedicated to the most elusive phenomenon in milling kinetics: non-linear breakage. Prof. Austin categorized this complicated phenomenon into two groups: "environment effects" and "material effects". The latter was treated by considering first-order breakage of soft and hard components or fast-breaking and slowly-breaking components of a mixture [10,11]. On the other hand, in general, the environment effects were found to be more elusive and complex. During ball milling of cement, e.g., Austin and Bagga [12] noted the slowing down of the breakage of all particle sizes and attributed this behavior to cushioning action of the fines. A non-first-order PBM based on the concept of warped time was developed to account for the cushioning phenomenon. To the best knowledge of the author, the approach to such elusive environmental effects was through the development of time-dependent specific breakage rate functions in PBMs until a true non-linear PBM was first developed in 2005 [13,14]. In fact, the so-called Model D of this non-linear theory [14] recovers Austin's approach as a limiting case (so-called uniform non-linear kinetics).

The last part of the talk is devoted to a brief discussion on recent advances in PBM parameter estimation [15–17] and the use of PBM–DEM (discrete element method) simulations [18,19] to gain mechanistic insights into the cushioning phenomenon. I demonstrate that DEM simulations shed light on the origin of such elusive breakage phenomenon like cushioning. Hence, the scientific journey into the realm of non-linear breakage, which was contributed significantly by Prof. Austin, continues. Overall, this talk highlights the key contributions of Prof. Austin's research to the modeling of milling processes within the context of PBM and their relevance and importance to current research in particle breakage and milling.

References

[1] L.G. Austin, A review: introduction to the mathematical description of grinding as a rate process, Powder Technol. 5 (1971), 1-17.

[2] L.G. Austin, R.R. Klimpel, P.T. Luckie, Process Engineering of Size Reduction: Ball Mill, Society of Mining Engineers of the AIME: Littleton, USA, 1984.

[3] B. Epstein, The mathematical description of certain breakage mechanisms leading to the logarithmico-normal distribution, J. Franklin Inst. 244 (1947) 471-477.

[4] B. Epstein, Logarithmico-normal distribution in breakage of solids, Ind. Eng. Chem. 40 (1948) 2289-2291.

[5] K. Sedlatschek, L. Bass, Contribution to the theory of milling processes, Powder Metal. Bull. 6 (1953) 148-153.

[6] L.G. Austin, P.T. Luckie, Methods for determination of breakage distribution parameters, Powder

Technol. 5 (1971/72) 215-222.

[7] L.G. Austin, V.K. Bhatia, Experimental methods for grinding studies in laboratory mills, Powder Technol. 5 (1971/72) 261-266.

[8] R.R. Klimpel, L.G. Austin, Determination of selection–for–breakage functions in the batch grinding equation by nonlinear optimization, Ind. Eng. Chem. Fund. 9 (1970) 230-237.

[9] R.R. Klimpel, L.G. Austin, The back-calculation of specific rates of breakage and non-normalized breakage distribution parameters from batch grinding data, Int. J. Miner. Process. 4 (1977) 7-32.

[10] L.G. Austin, T. Trimarchi, N.P. Weymont, An analysis of some cases of non-first-order breakage rates, Powder Technol. 17 (1977) 109–113.

[11] L.G. Austin, K. Shoji, D. Bell, Rate equations for non-linear breakage in mills due to material effects, Powder Technol. 31 (1982) 127–133.

[12] L.G. Austin, P. Bagga, An analysis of fine dry grinding in ball mills, Powder Technol. 28 (1981) 83–90.

[13] E. Bilgili, B. Scarlett, Population balance modeling of non-linear effects in milling processes, Powder Technol. 153 (2005) 59–71.

[14] E. Bilgili, J. Yepes, B. Scarlett, Formulation of a non-linear framework for population balance modeling of batch grinding: beyond first-order kinetics, Chem. Eng. Sci. 61 (2006) 33–44.

[15] M. Capece, E. Bilgili, R. Dave, Identification of the breakage rate and distribution parameters in a non-linear population balance model for batch milling, Powder Technol. 208 (2011) 195-204.

[16] E. Bilgili, M. Capece, A. Afolabi, Modeling of milling processes via DEM, PBM, and microhydrodynamics, in: P. Pandey, R. Bharadwaj (Eds.), Predictive Modeling of Pharmaceutical Unit Operations, Woodhead Publishing, Cambridge, 2017, pp. 159–203.

[17] N. Muanpaopong, R. Davé, E. Bilgili, A comparative analysis of steel and alumina balls in fine milling of cement clinker via PBM and DEM, Powder Technol. 421 (2023) 118454.

[18] M. Capece, R.N. Davé, E. Bilgili, A pseudo-coupled DEM–non-linear PBM approach for simulating the evolution of particle size during dry milling, Powder Technol. 323 (2018) 374-384.

[19] M. Capece, R.N. Davé, E. Bilgili, On the origin of non-linear breakage kinetics in dry milling, Powder Technol. 272 (2015) 189-203.

216b A Detailed Study of the Energy Utilisation in a Drop-Weight Tester

Ngonidzashe Chimwani, Mulenga Bwalya

Energy efficiency in comminution continues to be a prime concern as energy costs continue to escalate. Comminution is the largest consumer of energy in the mineral processing industry and that makes the quest for efficiency imperative. In that light, particle breakage has attracted attention over the years resulting in the invention of the drop-weight test machine among other important machines to measure the energy required to break a particle. However, accounting for the energy expenditure applied to the particle is still a challenge. Therefore, tests were conducted to study the breakage of silica sand under impact breakage, and account for energy dissipation in the form of heat using an infrared thermometer. Results show that energy expenditure during particle breakage can be determined from heat dissipation measured from the temperature changes. These findings provide important insights regarding energy usage during particle breakage and enable better control of the comminution process.

216c Scale-up Studies of the Spiral Air Jet Mill

Kunal Sharma, Devang Khakhar

The focus of the present work is the analysis and optimization of an air jet mill, which is a versatile equipment used in many industries for fine-grinding powders to sizes less than around 10 microns. The efficiency of the grinding process in an air jet mill typically varies from 2% to 4%, depending on the design and operating parameters so even small efficiency improvements can result in significant energy savings [1]. Amongst many of the available machines, spiral air jet mills are attracting attention because of their advantages like absence of the moving parts and low-temperature rise during grinding, resulting in better temperature control with minimal maintenance. Spiral air jet mills boast an exciting ability: not only can they grind feed particles, but they can also classify the resulting material as a product. This distinctive feature truly sets these machines apart.

From the impetus of the literature, as a first step, we study the gas flow in the mill using Computational Fluid Dynamics (CFD) with OpenFOAM and actual machine behavior through experiments. The spin number is the ratio of tangential velocity to the radial velocity of the flow. A particle inside the jet mill experiences the drag toward the outlet in the radially inwards direction and centrifugal force in the radially outward direction. A balance of the two forces defines a cut size of the particle at which both the forces are equal and opposite in direction. As per the resultant equation, the cut size is inversely proportional to the spin number, assuming other variables to be constant. Thus, as the system is dilute of particles, getting the flow behaviour of a single phase helps to predict the cut size variation.

The scaleup analysis of the spiral air jet mill involves an experimental study using various sizes of cylindrical mill chambers operated in a semi-batch mode. Initially, a fixed base case is established, where the diameter of the chamber is varied while maintaining a constant height. Subsequently, the height of the chamber is varied while keeping the diameter constant, similar to the base case scenario. During the experimental runs, the mill is fed in batches while the outlet is kept open to allow for continuous collection of the product over a specified duration. Each case is conducted under identical conditions to ensure consistency of the results. Following the experimental runs, both the mill material and the corresponding product undergo analysis. The size distribution of the particles is recorded, providing us with valuable statistical data for further analysis and interpretation.

The unimodal feed in the spiral air jet mill leads to a trimodal product, indicating the presence of multiple parallel grinding mechanisms within the mill. According to M.M. Dhakate et al. [2], the major contributors (by weight) to this trimodal distribution are erosion and chipping, followed by initial breakage and impact breakage. Enhancing the mill chamber in the radial direction has been shown to yield a finer mean product, with a similar trend observed in the leftover mill material. Conversely, varying the height of the mill chamber resulted in a finer product, but no clear trend was observed for the leftover mill material. It appears that the effect of radial variation dominates over the height variation of the mill chamber.

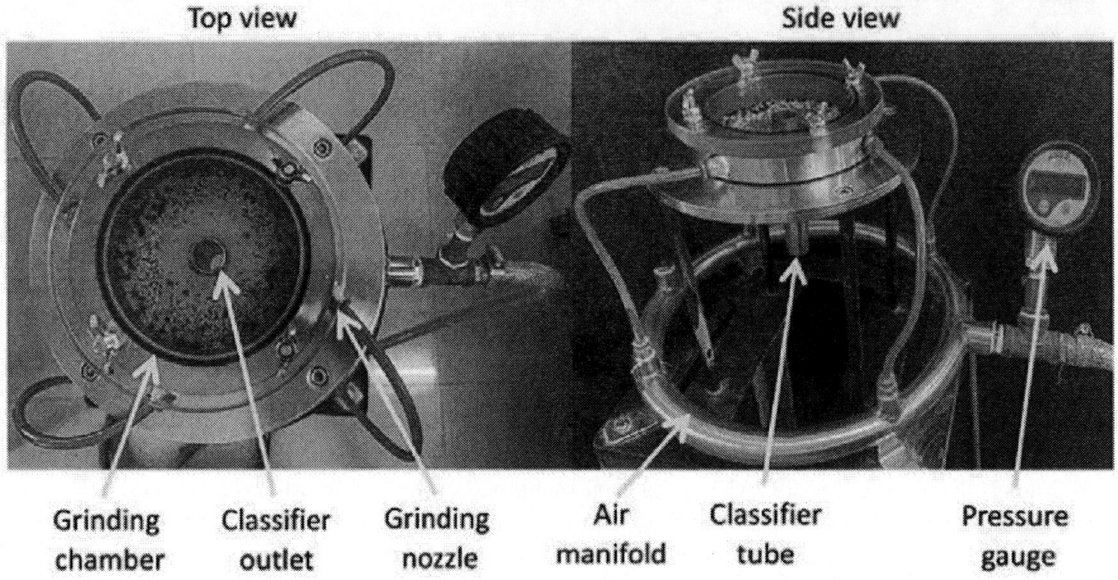

Photographs of the air jet mill used in the experiments with the different components labeled.

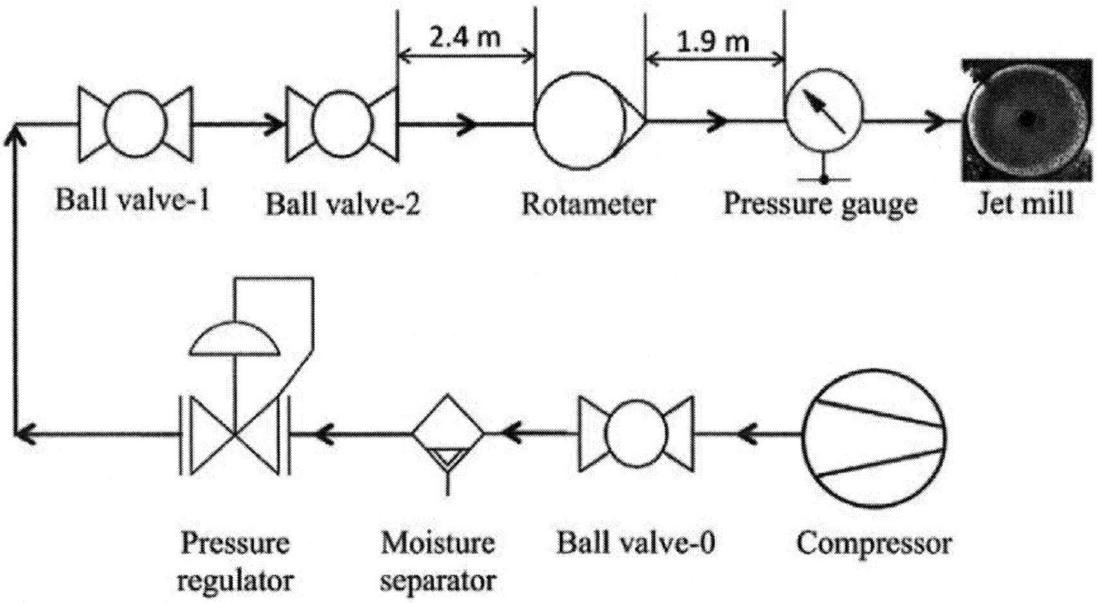

Schematic view of the experimental setup used in the experiments

216d Effect of Fines and Process Conditions on Particle Abrasion

Alejandro Gallegos Tovar, Lauren Endress, Wyatt Casey LaMarche, S. B. Reddy Karri, Jennifer Curtis

In flowing granular systems, such as conveying lines, fluidized beds, and hoppers, particles are subjected to different stresses which eventually lead to abrasion, or loss of small surface asperities. Abrasion impacts several key aspects of industrial operation, performance, and design, and often leads to environmental problems and the loss of valuable particle mass. In this study, particle abrasion is examined for three materials (pumice, bauxite, and FCC) in two different flowing granular systems. The first system is a FT4 shear cell in which the effects of increasing concentration of fines, as the particle abrades, and normal force are probed. When a normal force is applied at a constant shear rate, the generation of fines grows exponentially with time. However, if the fines in the shear cell are periodically removed, the abrasion rate is linear, indicating a cushioning effect of the fine particles on abrasion. The second system is a single-pass impact system in which a gas jet laden with particles impacts a flat plate, keeping the total solids loading for each pass constant. Single-pass systems allow for better assessment of the effect of fines, as the fines impact the plate the same number of times as the coarse material. In this system, the effect of fines concentration, gas velocity, particle size, and solids loading are investigated. Particle size distributions and microscopic images of the particles before and after abrasion, and high-speed video of particles in the near wall region, are also obtained. To isolate the effect of fines on particle abrasion, fine particles (under 44μm) are periodically removed after a given number of passes. Particle abrasion, as opposed to particle breakage, occurs below a critical gas velocity, and, without fines removal, the abrasion rate decreases as the number of passes increases.

216e Breakage of Structured Particles through Crack Propagation Phase-Field Model (PFM) Prediction Coupled with Population Balance Modeling

Kerry Johanson

Population balance models are powerful tools that relate the breakage observed to the probability that stress-stain events may induce the breakage of a given size particle. Population balance models relate the observed breakage to how a particular particle might break. They help determine how, or how much of, a particle of size bin 1 will break into size bin 2, size bin 3, size bin 4, etc. The trouble is that these population balance models are not predictive. They can analyze a set of batch breakage data and determine how and at what rate the various-sized particles might break, but they cannot look at a set of particles interacting and determine what particle shapes might result in the most probable breakage scenario. That scenario depends on the fracture pattern that any particle might have. The fracture pattern depends on the shape of the particle that breaks, how the particle interacts with other particles during stress-strain events, and the texture of the particle surface as well as the internal structure of the particle.

Consider an oval tablet made from five or more components. The fact that the particle is oval suggests that the particle may self-align with other oval particles to minimize the shear during a stress stain event. This limits the way that particle-particle interactions happen during shear. Next, consider what might happen if the particle shape were a series of crosses. As the series of cross-shaped particles interact, where would the preferred interaction happen and, more important, what stresses on this odd particle shape would induce a preferred breakage pattern? What if the particles are long rods – what would be the preferred breakage pattern? Or, suppose the particles are shaped like pretzels or curved potato chips with ridges. What if the particle of interest is a larger particle made of seeds of all sizes and shapes glued together with some form of binding agent? What if the particle of interest is a ceramic part designed with heat and cooling fins and a series of holes spaced in the particle so it can be attached to some form of electronic device? How would those particles interact with a similarly-shaped particle, and what would the preferred breakage planes be during stress-strain events?

Almost all materials subject to breakage have some form of structure associated with them. It is, therefore, reasonable to suggest that the breakage selectivity in the population balance model could be predicted from first principles if we could define a mathematical structure to capture the shape of the particles of interest and the internal structure of what makes up the particle. Here is where finite element and phase-field modeling come into play. The finite element method can compute the expected forces and stresses on any complex shape based on simple interactions on the surface of the particles. In a very simple particle example, consider a long slender rod-shaped particle that is supported between two contact points. As an additional particle presses on the rod it will break at the prescribed location where stresses are greatest predicted by FEM analysis. Very complex structures can be examined by finite element methods to predict stresses on the particle. A crack will begin to form when the local stresses approach the yield stress. However, the problem is that FEM analysis requires a continuum to compute the stress and strains, while the act of breakage induces gaps in this continuum, thereby making a FEM analysis difficult to do without complex and continual regeneration of meshes as new particles are fractured or as cracks grow in the particle of interest.

This complexity can be modeled using a phase-field model (PFM) to represent a crack geometry in a diffusive way without introducing sharp discontinuities. This enables PFM to model crack propagation using standard FEM. Thus, how a crack will form and propagate in a particle of a given configuration can be predicted. Applying a series of allowable contact points and positions to structured particles, and then using PFM to determine the location of the crack, allows engineers to develop a model that relates shape, powder properties, and potential particle-particle contact patterns to the breakage of structured particles. The powder flow properties help predict the breakage rates in the population balance model while the PFM crack propagation pattern helps predict the breakage stoichiometry. Thus, this approach allows the population balance model to be predictive from first principles.

216g Microhydrodynamic Scaling Analysis of a Vertical Wet Stirred Media Mill: Justifying PBM Parameters on Physical Grounds

Ecevit Bilgili, Alper Toprak, Deniz Altun, Okay Altun

Ore grades are decreasing due to the more complex form/mineralogy of the ore bodies over the years, which necessitates finer grind sizes to produce liberated minerals. According to various plant surveys [1,2], energy demand increases with product fineness exponentially. Therefore, selection of an appropriate grinding technology is important. Owing to their simple operation and design, conventional ball mills have been widely used, although they are known to be energy inefficient. The inefficiency of ball mills becomes especially obvious in <75 μm range [3]. Energy is imparted to the media by a rotating shell whose speed is in the range of 72%–80% of the critical speed, which limits the power input [4]. Wet stirred media milling (WSMM) has been used in the mining industry since 1980s; its use is on the rise due to higher demand for finer grinds [5]. In WSMM, a suspension of the ore particles is fed through a horizontal or vertical milling chamber which houses a rotor and beads upon filling. The high-speed rotation of the rotor induces turbulent flow and frequent collisions of the beads, which capture the particles in between and cause their compression–shear resulting in particle breakage. WSMM offers 30–40% energy saving compared to ball milling operated at the same size reduction ratio [6].

Modeling can help develop WSMM processes and provide quantitative insights into the impacts of process–design parameters [7]. The models include CFD–DEM [8], the stress intensity–stress number model [9], and the microhydrodynamic model [10,11]. Although they simulate the bead–bead collisions, particle capture efficiency, and their roles in WSMM with different levels of fidelity and complexity, they cannot simulate the evolution of the particle size distribution (PSD) in a transient process or predict the product PSD in a steady-state operation. To this end, population balance modeling (PBM) is used in which the particle size domain is discretized into N classes with sizes represented by x_i with $i = 1, 2,...N$ [12]. The PBM for milling describes how fast particles of size x_i break and where their broken pieces end up in the size domain or what the sizes of the progeny particles are. These quantitative descriptions are respectively provided by two fundamental breakage functions: the specific breakage rate function S_i and the cumulative breakage distribution function B_{ij}.

Many challenges exist when engineers are confronted with the inverse problem, i.e., estimation of S_i and B_{ij} from experimental data. The "*direct measurement methods*" include the one-size fraction method [13] and BI, BII, and BIII methods [14], respectively, and treat experimental batch ball milling data. They entail separate milling of multiple narrowly-sized feed samples covering the size range of interest; hence, they are experimentally laborious and time-consuming [12]. To mitigate some of these issues, the "*back-calculation method*" [15] has been used for the WSMM process [16], wherein a local optimizer along with the analytical or numerical solution of the PBM is used to estimate S_i and B_{ij} parameters simultaneously. The optimizer minimizes the sum of squared residuals (*SSR*) between the model predicted PSD and the experimental PSD. However, a major challenge since the early applications of the back-calculation method is to find a unique set of the PBM parameters [12,15]. A local optimizer could get stuck at one of the local minima, leading to an erroneous set of parameters despite yielding a deceptively reasonable fit to the PSD data. The non-uniqueness of the solution is revealed when the set of initial guesses for the parameters to be estimated is varied, which leads to different parameter estimates [17]. Therefore, the use of a global optimization scheme in the back-calculation with PBMs is warranted.

Not only should an industrially relevant PBM describe and fit existing plant data, but also it should be able to predict the product PSD under different processing conditions for different feed PSDs. Moreover, its S_i and B_{ij} parameters must be sufficiently accurate and physically plausible to inform engineers about the breakage mechanisms and dependence of S_i on the process–design parameters. On the other hand, such a PBM could have up to a dozen S_i and B_{ij} parameters to be estimated, and even the use of a global optimizer on a sufficiently dense data set cannot guarantee the accuracy and the uniqueness of the estimated parameters. There are two possible ways to resolve this challenge: (i) reduce the number of parameters using a hybrid approach and/or (ii) use modeling approaches that are more mechanistic along with PBMs as well as physical plausibility arguments to justify the parameters estimated. In the hybrid approach (i), a direct measurement method is used first to estimate the B_{ij} parameters and then the so-obtained B_{ij} was input to the back-calculation method to estimate the S_i parameters solely [18]. This presentation focuses on the second approach (ii).

Using a microhydrodynamic scaling analysis, we aimed to justify the S_i parameters of a PBM, which was used to model an industrial-scale vertical wet stirred media mill. Feed and product particle size distributions (PSDs) of copper ore were measured at various steady-state conditions. Starting with the microhydrodynamic theory [11] and the definition of S_i for impact breakage [19], a simple power-law relation between S_i and the rotation rate of the rotor (ω), % ore content (w_p), and particle size (x_i) was derived. The impact of the suspension volumetric flow rate Q_S was accounted for via the mean residence of the PBM. The PBM incorporated 1-large–2-small-tanks model and the power-law specific breakage rate function S_i. Parameters of S_i and a non-normalized cumulative breakage function B_{ij} were estimated by a back-calculation method with global optimization. Fig. 1 shows the product PSDs in 18 different processing conditions fitted reasonably well by the PBM. Results suggest that higher rotor speed, lower solids loading, and use of steel rotor vs. rubber rotor and coarser beads led to faster breakage. The exponents revealed by the microhydrodynamic scaling analysis were within 17% of the exponents estimated by the back-calculation. Overall, we have demonstrated the use of the microhydrodynamic theory to justify the physical plausibility of the PBM parameters.

References

[1] F. Shi, R. Morrison, A. Cervellin, F. Burns, F. Musa, Comparison of energy efficiency between ball mills and stirred mills in coarse grinding, Miner. Eng. 22 (2009) 673–680.

[2] L. Little, A. N. Mainza, M. Becker, J. Wiese, Fine grinding: How mill type affects particle shape characteristics and mineral liberation. Miner. Eng. 111 (2017) 148–157.

[3] A. Jankovic, Variables affecting the fine grinding of minerals using stirred mills. Miner. Eng. 16 (2003) 337–345.

[4] A. Giblett, B. Putland, Grinding circuit design, In SME Mineral Processing & Extractive Metallurgy Handbook; Dunne. R.C., Kawatra, K., Young, C.A., Eds.; SME: Littleton, CO, USA, 2019.

[5] L. Taylor, D. Skuse, S. Blackburn, R. Greenwood, Stirred media mills in the mining industry: Material grindability, energy–size relationships, and operating conditions. Powder Technol. 369 (2020) 1–16.

[6] A. Gupta, D. Yan, Mineral Processing Design and Operations: An Introduction, 2nd Ed., Elsevier, 2016.

[7] E. Bilgili, G. Guner, Mechanistic modeling of wet stirred media milling for production of drug nanosuspensions, AAPS Pharm. Sci. Technol. 22 (2020) 2.

[8] C.T. Jayasundara, R.Y. Yang, B.Y. Guo, A.B. Yu, Govender I, Mainza A, et al. CFD–DEM modelling of particle flow in IsaMills–Comparison between simulations and PEPT measurements, Miner. Eng. 24 (2011) 181–187.

[9] A. Kwade, Determination of the most important grinding mechanism in stirred media mills by calculating stress intensity and stress number, Powder Technol. 105 (1999) 382–388.

[10] D. Eskin, O. Zhupanska, R. Hamey, B. Moudgil, B. Scarlett, Microhydrodynamic analysis of nanogrinding in stirred media mills, AIChE Journal. 51 (2005) 1346–1358.

[11] G. Guner, D. Yilmaz, D. Eskin, E. Bilgili, Effects of bead packing limit concentration on microhydrodynamics– based prediction of breakage kinetics in wet stirred media milling, Powder Technol. 403 (2022) S117433.

[12] L.G. Austin, R.P. Klimpel, P.T. Luckie, Process engineering of size reduction: Ball milling. New York, NY: Society of Mining Engineers of the AIME, 1984.

[13] L.G. Austin, V.K. Bhatia, Experimental methods for grinding studies in laboratory mills, Powder Technol. 5, (1971–72) 261–266.

[14] L.G. Austin, P.T. Luckie, Methods for determination of breakage distribution parameters, Powder Technol. 5 (1971–72) 215–222.

[15] R.R. Klimpel, L.G. Austin, The back– calculation of specific rates of breakage and non– normalized breakage distribution parameters from batch grinding data, Int. J. Miner. Process. 4 (1977) 7–32.

[16] E. Bilgili, R. Hamey, B. Scarlett, Nano– milling of pigment agglomerates using a wet stirred media mill: Elucidation of the kinetics and breakage mechanisms, Chem. Eng. Sci. 61 (2006) 149–157.

[17] N. Muanpaopong, R. Davé, E. Bilgili, A comparative analysis of steel and alumina balls in finemilling of cement clinker via PBM and DEM, Powder Technol. 421 (2023) 118454.

[18] Y.M. Zhang, A. Kavetsky, Investigation of particle breakage mechanisms in a batch ball mill using back– calculation, Int. J. Miner. Process. 39 (1993) 41–60.

[19] M. Capece, E. Bilgili, R.N. Davé, Formulation of a physically motivated specific breakage rate parameter for ball milling via the discrete element method, AIChE J. 60 (2014) 2404–2415.

Acknowledgement: This work was equally contributed by all authors.

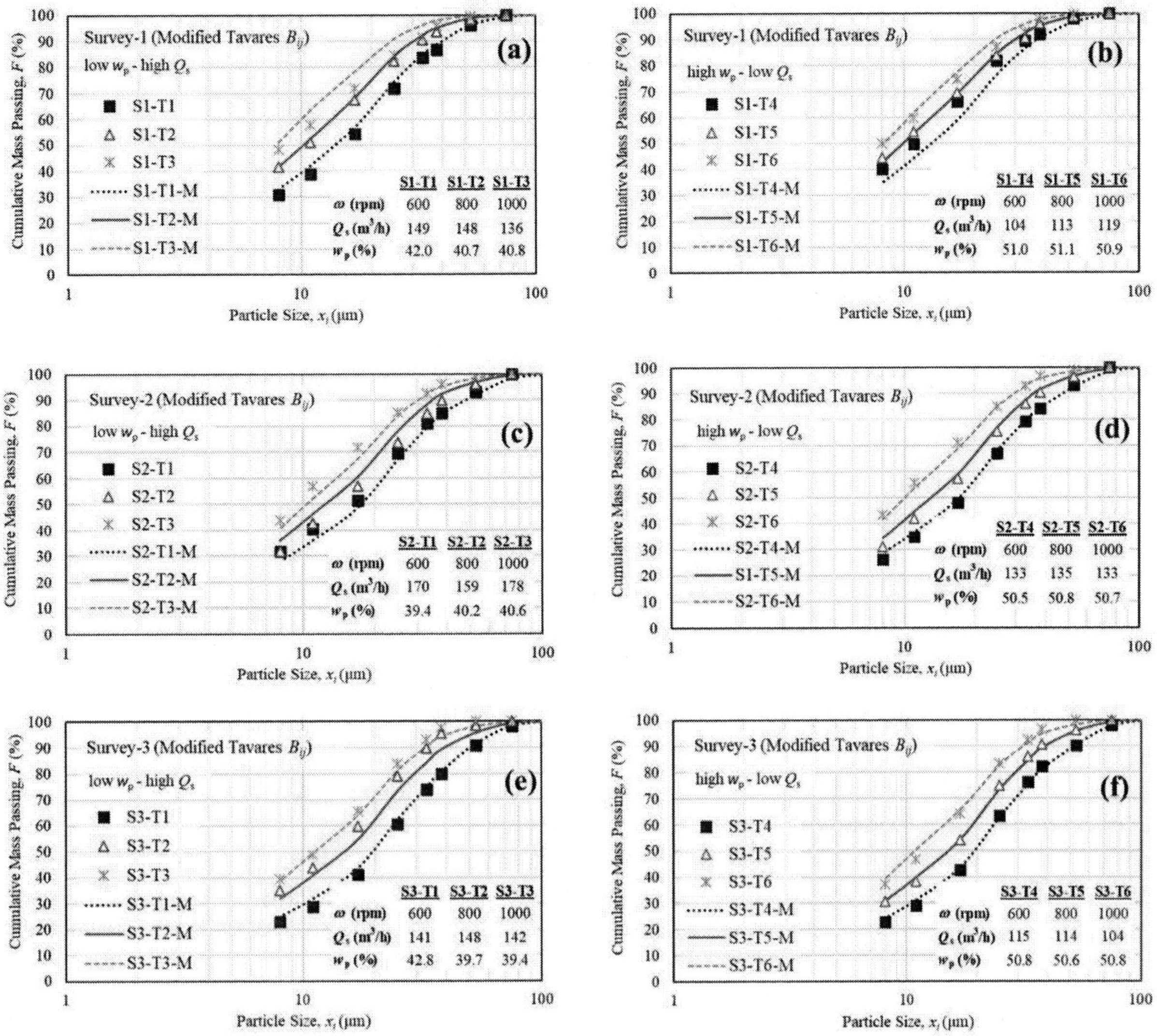

Fig. 1. Fits of the cumulative product PSDs using the back-calculation method with the modified Tavares B_{ij}: (a) Survey 1 with low w_p–high Q_s, (b) Survey 1 with high w_p–low Q_s, (c) Survey 2 with low w_p–high Q_s, (d) Survey 2 with high w_p–low Q_s, (e) Survey 3 with low w_p–high Q_s, and (f) Survey 3 with high w_p–low Q_s. Abbreviations: ω: rotation rate of the motor, Q_s: suspension volumetric flow rate, and w_p: % ore loading.

148a Characterisation of Eco-Friendly Calcium Carbonate-Polydopamine Microcapsules with Superior Mechanical, Barrier, and Adhesive Properties

Daniele Baiocco, Benjamin Lobel, Mohammed Al-Sharabi, Alexander Routh, Olivier J. Cayre, Zhibing Zhang

Introduction: There is a rising need to effectively deliver perfume molecules onto fabric surfaces during laundry washing to enhance consumers' satisfaction. To this end, core-shell microcapsules fabricated with synthetic polymers (e.g. melamine-formaldehyde) have long played a pivotal role in detergent manufacturing industry. Despite their excellent performance, synthetic polymers are non-biodegradable and accumulate as microplastics in the environment [1]. Consequently, the development of novel eco-friendly microcapsules is imperative, without undermining their performance. In the literature, inorganic materials have proven effective to form microcapsule shells [2]. Indeed, calcium carbonate (CC) is environmentally benign and can yield robust, cost-effective microcapsules. However, scarce information on the performance of the ensuing microcapsules is available, possibly suggesting that their barrier and adhesive properties were unsatisfactory [3]. Bioinspired by the extremely adhesive properties of catecholamines (CA) in mussels, dopamine is a promising CA-rich building block monomer. In its polymerised form, namely polydopamine (PDA), it can form a homogeneous layer onto many surfaces [4].

Methods: This work aims to develop core-shell hybrid microcapsules with a core of model perfume oil hexyl salicylate (HS) within a solid inorganic (CC) shell *via* interfacial crystal ripening followed by the deposition of an organic coating *via* oxidative alkaline auto-polymerisation of dopamine (pH 8.5). Their morphology, mechanical and barrier properties were characterised *via* fluorescence sensing/scanning electron microscopy (SEM), a micromanipulation technique, and UV-Vis spectrophotometry, respectively. Moreover, a bespoke microfluidic device was employed to investigate the retention performance of PDA-coated microcapsules onto a model fabric substrate (polyethylene terephthalate (PET)) in washing-mimicked conditions.

Results: There appeared to be mostly spherical core-shell microcapsules ($D_{[3,2]}$=31.2±0.4 μm) with a smooth PDA-coated surface beneath which CC was porously formed. After one month in water, only ~40% of HS was released, whereas PDA-free CC microcapsules had released the whole HS load within <4 hours. Under compression, the microcapsules yielded a mean nominal rupture tension of 51.8±6.2 N/m, which is similar to commercial microcapsules [5]. At pH 9 of a typical detergent, their retention performance on PET films was significantly high (~60%) under washing-like hydrodynamic shear stress (1 Pa), which may be appealing to industries [1].

Conclusions: The results suggest that hazard-free perfume microcapsules with a hybrid shell can be fabricated effectively, leading to superior adhesive properties and sustained release, with potential applications in multiple fast-moving consumer goods.

Learning objectives: Attendees will identify new tools for characterising and enhancing the key performance properties of microcapsules. Participants will be able to appraise the potential of microplastic-free hybrid microcapsules for potential controlled release applications.

Acknowledgments

This work was supported by EPSRC, UK (Grant Number EP/V027654/1).

References

1. Baiocco, D. and Z. Zhang, *Microplastic-Free Microcapsules to Encapsulate Health-Promoting Limonene Oil.* Molecules, 2022. **27**
2. Yeom, J., W.S. Shim, and N.G. Kang *Eco-Friendly Silica Microcapsules with Improved Fragrance Retention*. Applied Sciences, 2022. **12**, DOI: 10.3390/app12136759.
3. Wang, X., W. Zhou, J. Cao, W. Liu, and S. Zhu, *Preparation of core–shell CaCO3 capsules via Pickering emulsion templates.* Journal of Colloid and Interface Science, 2012. **372**(1): p. 24-31.
4. Kang, S., M. Baginska, S.R. White, and N.R. Sottos, *Core–Shell Polymeric Microcapsules with Superior Thermal and Solvent Stability.* ACS Applied Materials & Interfaces, 2015. **7**(20): p. 10952-10956.
5. Long, Y., B. Vincent, D. York, Z. Zhang, and J.A. Preece, *Organic-inorganic double shell composite microcapsules.* Chem Commun (Camb), 2010. **46**(10): p. 1718-20.

148b Comprehensive Powder Flow Characterization with Reduced Testing

Catarina Chendo, João F. Pinto, Maria Paisana

Powder flow is a critical attribute of pharmaceutical blends to ensure tablet weight uniformity and production of tablets with consistent and reproducible properties. This study aims at characterizing different powder blends with a number of different rheological methodologies, in order to understand how particle physical attributes (namely morphology and size) and interaction between components within the formulation generate different responses when analysed by different rheological tests. Furthermore, this study intends on reducing the number of rheological tests in early development phases, by focusing on the tests that generate the maximum information about the flowability attributes of the pharmaceutical blends, avoiding redundant characterization. This work considered two cohesive powders - spray-dried hydroxypropyl cellulose (SD HPMC) and micronized indomethacin (IND) – formulated with other four commonly used excipients [(lactose monohydrated (LAC), microcrystalline cellulose (MCC), magnesium stearate (MgSt) and colloidal silica (CS)]. The experimental results showed that powder flowability may be affected by materials particles' size, bulk density, morphology, and interactions with lubricant. In detail, parameters, such as angle of repose (AoR), compressibility percentage (CPS), and *flow function coefficient* (ff_c) have shown to be highly affected by the particle size of the materials present in the blends. On the other hand, the Specific Energy (SE) and the effective angle of internal friction (φe) showed to be more related with particle morphology and materials interaction with the lubricant. Since both ff_c and φe parameters are generated from the same *yield locus* test, data suggest that a number of different powder flow features may be understood only by applying this test, avoiding redundant powder flow characterization, as well as extensive time and material spent in early development formulation stages.

148c Multidimensional Nanoparticle Characterization

Wolfgang Peukert, Johannes Walter

An important trend in particle science and technology is the requirement to understand and design systems of ever increasing complexity. Available techniques for particle characterization are predominantly focused on size characterization, even simultaneous size and shape characterization is a major challenge, in particular for nanoparticles. Property functions as a comprehensive description of such particle systems must address both, single particles and the structure of their assemblies. The property function defines the target in product design and is understood along the five dimensions of particle size, particle shape, particle surface, particle structure (i.e. amorphous, crystalline, internal pore size distributions and defects), particle composition and their respective distributions. While the characterization of 1D particle size distributions is well established, the independent description and measurement of two or more distributed properties of a particle ensemble is a highly challenging task. Recent progress in the description and characterization of multivariate particle property distributions will be presented and discussed [1].

The typical way of particle characterization via imaging technologies like optical image analysis, SEM or TEM allow access to the 2D projections of particles. AFM imaging together with statistical RAMAN mapping can provide information on particle dimensions together with information on the complex chemical composition. Most of these techniques, in particular those for submicron particles, are very time-consuming, require complex sample handling and share the common problem of poor statistics. The obtained information is based on a limited number of counts and does not necessarily contain sufficient particles to approximate the underlying nD distributions properly.

Progress in analytical ultracentrifugation (AUC) combines outstanding resolution with high reproducibility down to particles of 1 nm in size. Novel sensors based on absorption and emission optics [2] provide access to the optical properties of size-selected nanoparticles. This allows accurate the size measurement of multimodal samples due to the excellent separation power in strong centrifugal fields. In principle, all particle systems with size, shape or compositional, respectively density, dependency of sedimenation and optical properties can be analyzed avoiding tedious TEM- or AFM-based counting. AUC was employed for the efficient characterization 2D materials such as graphene oxide or MoS_2 giving fast access during process development. Full 2D characterization of core-shell nanoparticles by the analysis of sedimentation profiles based on sedimentation and diffusion properties are demonstrated and validated by combined SAXS/SANS measurements. This enables, for instance, the characterization of size-band gap dependencies of quantum dots. Recent approaches based on size-dependent optical properties of plasmonic nanoparticles allow to measure full number-weighted 2D distributions of size and length distributions of gold nanorods or of 5-sided gold double pyramids in one single experiment [3,4]. In a similar approach, size and compositional distributions of noble metal alloy nanoparticles can be obtained [5]. Analytical ultracentrifugation is considered as gold standard in nanoparticle characterization due to the unprecedented resolution, accuracy and reproducibility.

Funding by DFG of the Collaborative Research Center 1411 "Design of Particulate Products" is greatfully acknowledged (https://www.crc1411.research.fau.eu/).

References

[1] Frank U., Uttinger M.J. Wawra S.E., Lübbert C., Peukert W., Progress in multidimensional

particle characterization, KONA (invited review) 39 (2022) 3-28

[2] Lautenbach V., Onishchukov, G., Wawra S., Frank U., Hartmann L., Peukert W., Walter J., Development of an advanced multiwavelength emission detector for the analytical ultracentrifuge, Nanoscale Advances (2024), in press

[3] Wawra S., Thajudeen T., Pflug L., Kryschi C., Stingl M., Peukert W., Determination of the two-dimensional distributions of gold nanorods by multiwavelength analytical ultracentrifugation, Nature Communications 9 (2018) 4898

[4] Frank U., Drobek D., Sanchez-Iglesias A., Wawra S., Nees N., Walter J., Pflug L., Apeleo Zubiri B., Spiecker E., Liz-Marzan L., Peukert W., Determination of 2D particle size distributions in plasmonic nanoparticle colloids via analytical ultracentrifugation - Application to gold bipyramids, ACS Nano 17 (2023) 5785-5798

[5] Cardenas Lopez, P., Uttinger M.J., Traore N.E., Drobek D., Apeleo Zubiri B., Spiecker E., Peukert W., Walter J., Multidimensional characterization of noble metal alloy nanoparticles by multi-wavelength analytical ultracentrifugation, Nanoscale 9 (2022) 14, 12928-12939

148d Scalable Manufacturing of X-Ray Compatible Microfluidics for High-Throughput Structure Determination and Integrated Liquid Handling Strategies

Sarthak Saha, Yaozu Chen, Sarah Perry

Silicone-based elastomers, along with thermoplastics and paper, represent the most widely used materials for microfluidic devices. However, applications in structural biology require a level of X-ray transparency that is difficult to achieve with traditional materials. A key aspect of protein structure determination is the manual harvesting and freezing of protein crystals, which are analyzed by X-ray diffraction to resolve the final 3D structure. Here we demonstrate the use of UV-curable polymers that have good X-ray transparency and further allow for automated sample loading towards high-throughput and/or remote X-ray data collection. We additionally showcase continuous manufacturing of these devices to enable more cost-effective large-scale manufacturing that is 80 times faster than traditional batch fabrication strategies and thus reduce labor costs significantly.

We leverage this approach to develop microfluidic devices that allow for on-chip crystallization and structure determination. The reagent manipulation requirements for crystallization experiments calls for precise liquid handling strategies and therefore incorporation of a novel valving mechanism compatible within the UV-curable device architecture. Our design incorporates advanced fluid handling systems, realized through both centrifugal and electrically actuated valves, departing from conventional pressure-driven systems typically associated with cumbersome equipment like pneumatic controllers and nitrogen cylinders. Our devices provide a one-step solution for structural biologists, from screening crystallization conditions to determining the protein structure. These capabilities in a microfluidic device also have the potential to enable next generation structural biology experiments, like ligand binding and time-resolved experiments, while bringing us one step closer to automating the labor-intense process of crystallization and crystallography.

148e A Novel Technique to Characterize the Phenomena of Crystallization and Aggregation in a Suspension Settling in a Vertical Tube

Rutuja Amrale, Jyoti Seth, Vinay Juvekar

Studies of dynamic processes like crystallization, nanoparticle synthesis, and particle-particle aggregation require accurate and instantaneous measurements of particle size distribution. Existing methods, like Dynamic Light Scattering (DLS), Dynamic Image Analysis (DIA), and Laser Diffraction, measure particle size distributions in static solutions, have to be carried out periodically, and are susceptible to sampling errors. We address the need for dynamic and in-situ measurements of particle size distribution using a combination of dynamic distribution of particles using static multiple light scattering and Stokes' law for particle settling using a population balance model.

In our experimental setup, particles are allowed to settle in a vial under gravity, and the fraction of incident light transmitted through the suspension is measured along the length of the vertical cell through which the particles settle (Turbiscan LAB, Formulaction make). The data is then modelled using the population balance model, which accounts for particle growth and adsorption coupled with settling, and also the variation of solute concentration in the continuous phase. The settling velocity of the particle depends on the diameter of the particle, while the transmitted intensity is a function of the particle volume fraction and the mean particle diameter. The model equations are solved by the method of characteristics matching the transmission wave profiles. By monitoring the change in the transmission profile over time, particle size distribution in a suspension can be obtained.

We have demonstrated this technique for systems undergoing the processes of crystallization and aggregation. During crystallization, nucleation and growth of particles increase the volume fraction and mean particle diameters, which are estimated by observing the settling profiles of the suspension, and an instantaneous particle size distribution is obtained. During aggregation, the particle sizes and densities change, which again influences the settling profiles of the suspension. We have studied the formation of calcium carbonate crystals and the role of polymers in modifying the growth and/or aggregation characteristics. Independently, to study the process of aggregation, we consider the aggregation of silica particles on cellulose particles and the influence of polymers on the adsorption of silica particles on cellulose.

This versatile technique can be used to analyse complex phenomena involving particle nucleation and growth and particle-particle interactions, without removing the sample from the reservoir.

148f Role of *Staphylococcus Aureus*'s Buoyant Density in the Development of Biofilm Associated Antibiotic Susceptibility

Sarah Kispert, Madison Liguori, Cody Velikaneye, Chong Qiu, Shue Wang, Nan Zhang, Huan Gu

Bacterial biofilms are clusters of bacterial cells that form at various interfaces, including those between air and liquid or liquid and solid. Due to their roles in enhancing wastewater treatment processes, and their unfortunate propensity to cause persistent human infections through lowering antibiotic susceptibility, understanding and managing bacterial biofilms is of paramount importance. A pivotal stage in biofilm development is the initial bacterial attachment to these interfaces. However, the determinants of bacterial cell choice in colonizing an interface first and heterogeneity in bacterial adhesion remain elusive. Recent research has unveiled variations in the buoyant density of biological particles such as *Staphylococcus aureus* cells, irrespective of their growth phase. Cells with a low cell buoyant density, characterized by fewer cell contents, exhibited lower susceptibility to antibiotic treatments (100 μg/mL vancomycin) and favored biofilm formation at air-liquid interfaces. In contrast, cells with higher buoyant cell density, which have richer cell contents, were more vulnerable to antibiotics and predominantly formed biofilms on liquid-solid interfaces when contained upright. In essence, *S. aureus* cells with higher buoyant cell density may be more inclined to adhere to upright substrates.

37a A Novel Hypergravity Platform to Tune the Particle Size Distribution of Small Molecules with Increasing g-Levels

Adrian Radocea

Microgravity can have a meaningful impact on pharmaceutical research and development. Decades of research on the International Space Station (ISS) have shown that a reduced gravity environment can enable improved control over particle size, shape, and crystal quality[1], which can unlock new therapeutic formulations. Commercialization of reusable rockets has increased access to low-Earth orbit and provides opportunities to conduct pharmaceutical R&D in space at relatively low cost. Microgravity suppresses convection and sedimentation, and the resulting diffusion-driven transport changes the crystallization kinetics. In contrast, an increased gravity environment achieved through centrifugal forces changes nucleation and growth rates by changing local concentration gradients. In this work[2], a large diameter centrifuge was used to generate high centrifugal forces. A Technobis Crystal16 reactor[3] was strapped on to an arm of the centrifuge to carry out seeded cooling crystallizations of L-histidine at variable effective g-forces. It was shown that changing g-levels has a nonlinear effect on the particle size distribution of the product crystals. With increasing g-levels, the mode of the PSD initially increases, however further increases in g-level result in a reversal of the trend, and the mode of the PSD is reduced. State-of-the-art process analytical technology (PAT) including in-situ microscopy, and in-situ Raman spectroscopy were used to track the state of the system during the crystallization process. A hypothesis was constructed which depicted a spatial variation of supersaturation with increasing g-levels, which could explain the shift in the mode of the PSD with increasing g-levels. This hypothesis was validated with carefully designed dissolution experiments. The role of crystal breakage as a possible mechanism which can explain the shifts of the PSD was refuted based on experiments carried out at different g-levels. Our novel hypergravity platform can play a crucial role in generating gravity-sensitivity data for molecules to help down-select promising molecules and process conditions for microgravity flight studies.

References

1. Snell, E. H., & Helliwell, J. R., Microgravity as an environment for macromolecular crystallization–an outlook in the era of space stations and commercial space flight. Crystallography Reviews, **2021**, 27(1), 3-46.
2. Pal K., Radocea A., Gravity as a knob for tuning particle size distributions of small molecules, *Crystal Growth and Design*, **2024**, 24(6), 2301-2666.
3. https://www.crystallizationsystems.com/products/crystal16/

37b Insights into Particle-Droplet Collisions

Andrew Bayly, Tushae Srivastava

Collisions between particles and droplets are important to the performance of spray dryers and many other spray processes. The dynamics and outcome of these collisions depends on the particle and droplet properties and their initial velocities. The work presented here focuses on collisions where the particle is smaller than the droplet, as s is common in spray dryers. It experimentally explores the influence of the particle and droplet properties on the dynamics of the collision and their influence on the collision outcome. It uses imaging and image analysis to investigate free binary collisions where droplet rotation is observed to be important and highlights the significant impact of contact angle and density on the collisions. Thus providing useful insights and empirical models that have the potential to be used in the simulation of spray processes.

37c Impacts of Surface Composition on Dissolution Performance of Amorphous Solid Dispersions

Yingzhen Ma, Pavithra Sundararajan, Luke Schenck, Ashish Punia, Jasmine Rowe, Daniel Smith

Amorphous solid dispersion (ASD) plays a governing role in improving the bioavailability and solubility of poorly soluble drugs. Amorphous solid dispersions are a solid solution of drug, polymer and sometimes even surfactants. Depending on the phase behavior and miscibility of the components of ASD, process variations can sometimes lead to compositional differences on the surface. In this work, we study and understand how changes in surface concentration of ASD with same composition can impact their physical stability and dissolution behavior. Spraying drying technique is utilized in our experiments to generate ASD. We employ a formulation containing polyvinyl pyrrolidone vinyl acetate (PVP-VA), sodium lauryl sulfate (SLS) and drug. Surface concentration can be modified by varying the solvent systems and solid loading used in the spray drying process while maintaining the same ASD composition. Dissolution and accelerated physical stability of the resulting ASD were assessed and compared to a reference process. The ASD samples with different dissolution performance / physical stability were further evaluated for differences in surface concentration. We have thus confirmed that surface concentration of ASD can be influenced by process changes while preserving the identical ASD composition. We will explore the driving forces that lead to differences in surface concentration and gain fundamental insights into interplay of these mechanisms on the drug product performance.

37d Environmentally and User-Friendly Composite Microcapsules with Superior Mechanical Properties for Potential Applications in Fast-Moving Consumer Goods

Daniele Baiocco, Mohammed Al-Sharabi, Benjamin Lobel, Alexander Routh, Olivier J. Cayre, Zhibing Zhang

Core-shell microcapsules are used in the broader fast-moving consumer goods (FMCG) industry to encapsulate value-added fragrances, flavours, or other bioactive ingredients. These microcapsules encase functional species within protective shells, thereby enabling their controlled and efficient release. This leads to enhanced product effectiveness and prolonged shelf-life of the final FMCG, all while safeguarding costly ingredients from harsh environmental factors (temperature, pH, UV radiation) [1]. In recent years, the mechanical properties of microcapsules have garnered significant attention due to their direct influence on stability, durability, and overall performance in a plethora of end-use applications, such as detergents, foams, and everyday cosmetics. To this end, various shell materials are engineered to meet specific mechanical requirements [2]. Interestingly, synthetic shells, such as polyurethane, polyurea and polystyrene, can offer a balance of thermal, chemical and mechanical resistance, as well as providing protective barrier for active species. However, they are largely non-biodegradable and contribute to the build-up of microplastics in the environment [3]. Inorganic shells like silica can provide high thermal stability while remaining chemically inert. However, silica is inherently brittle, making silica-based shells subject to cracking or breaking under mechanical stress. Calcium carbonate (CC) is environmentally benign and has been demonstrated to economically form microcapsule shells [4]. Despite its great potential, their mechanical and barrier properties remain to be significantly improved for multifunctional applications.

Herein we present core-shell composite microcapsules with an inert model hydrophobic core (hexyl salicylate) which had been stabilised both electrostatically and sterically utilising cetyltrimethylammonium bromide (CTAB) and fumed hydrophilic silica nanoparticles (SiO_2 NPs), respectively. A seeded CC shell was formed via interfacial crystal ripening facilitated by the presence of polyacrylic acid (PAA) [5]. Subsequently, an additional coating of polydopamine (PDA) was formed *via* oxidative auto-polymerisation (pH 8.5) to enhance the structural stability of the CC crystals. The morphology and barrier properties of the ensuing microcapsules were characterised *via* fluorescence sensing/scanning (SEM)/transmission electron microscopy (TEM), energy dispersive x-ray (EDX) analysis, and UV-Vis spectrophotometry, respectively. Importantly, their mechanical properties were assessed by a micromanipulation technique based on parallel plate compression of single particles [6].

There appeared to be core-shell microcapsules ($D_{[3,2]} = 19.7 \pm 0.3$ μm) with a spherical morphology. Topographical analysis revealed a continuous and relatively smooth surface (Figure 1). EDX analyses confirmed the presence of SiO_2 (1.0% wt.) due to the nanoparticles and a high content of CC (15.2% wt.). Over 24 h in a receptor medium (36% (v/v) hydro-propanol) where the solubility of hexylsalicylate is ~10.5 kg/m^3 at 25 °C, ~50% of the core was released, with the remaining load being released within 6 days. The corresponding shell permeability/shell thickness was evaluated to be ((5.8 $\pm$ 1.3)$\cdot 10^{-7}$ m/s) via solute-diffusion modelling, which is statistically similar to that of microcapsules with a shell of melamine formaldehyde ((2.4 $\pm$ 0.9)$\cdot 10^{-6}$ m/s) [7]. TEM analysis confirmed the presence of a multi-layer composite shell, featuring a thick CC based coating of ~1 μm. Under compression, the microcapsules yielded a mean nominal rupture stress and tension of 73.5 $\pm$ 5.0 MPa and 485.6 ± 28.8 N/m, respectively, which are significantly greater than any microcapsules reported in literature. This finding confirms the extremely high mechanical strength of these microcapsules, which could be particularly appealing to many industrial sectors, including pharmaceuticals, nutraceuticals, personal care, cosmeceuticals, to name a few. Details of the study will be presented.

References

1. Baiocco, D., J.A. Preece, and Z. Zhang, *Encapsulation of Hexylsalicylate in an Animal-Free Chitosan-Gum Arabic Shell by Complex Coacervation.* Colloids and Surfaces A: Physicochemical and Engineering Aspects, 2021: p. 126861.
2. Mercadé-Prieto, R. and Z. Zhang, *Mechanical characterization of microspheres – capsules, cells and beads: a review.* Journal of Microencapsulation, 2012. **29**(3): p. 277-285.
3. Baiocco, D. and Z. Zhang, *Microplastic-Free Microcapsules to Encapsulate Health-Promoting Limonene Oil.* Molecules, 2022. **27**.
4. Long, Y., B. Vincent, D. York, Z. Zhang, and J.A. Preece, *Organic-inorganic double shell composite microcapsules.* Chem Commun (Camb), 2010. **46**(10): p. 1718-20.
5. Donnet, M., A. Aimable, J. Lemaître, and P. Bowen, *Contribution of Aggregation to the Growth Mechanism of Seeded Calcium Carbonate Precipitation in the Presence of Polyacrylic Acid.* The Journal of Physical Chemistry B, 2010. **114**(37): p. 12058-12067.
6. Zhang, Z., R. Saunders, and C.R. Thomas, *Mechanical strength of single microcapsules determined by a novel micromanipulation technique.* Journal of microencapsulation, 1999. **16**(1): p. 117-124.
7. Mercade-Prieto, R., R. Allen, D. York, J.A. Preece, T.E. Goodwin, and Z. Zhang, *Determination of the shell permeability of microcapsules with a core of oil-based active ingredient.* J Microencapsul, 2012. **29**(5): p. 463-74.

37e Aggregation and Breakage Dynamics of Alumina Particles Under Laminar Shear By CFD-DEM Simulations

Lequan Zeng, George Franks, Eirini Goudeli

Particle recovery from suspension holds significant importance across various industries including wastewater treatment, food engineering, ceramic powder processing, and mineral processing. An illustrative application is the aggregation of fine valuable mineral particles in mineral processing to facilitate their collection via froth flotation. Thus, effective control of particle aggregation and breakage dynamics is key in powder and particle processing, given that the aggregate characteristics influence product grade and process efficiency [1]. Various control strategies can be employed to manipulate interparticle interactions, including adjusting suspension pH, using salts, adding surface modifiers like polymers, or applying shear forces to suspensions [2]. However, observing and quantifying particle dynamics in suspensions experimentally poses challenges due to the rapid particle aggregation and breakage, usually taking place within microseconds. Particle-based simulations provide a promising approach for studying particle dynamics, enabling direct observation and quantification of aggregate formation and breakage in suspensions [1].

Here, Computational Fluid Dynamics – Discrete Element Method (CFD-DEM) are used to explore how laminar shear affects particle aggregation and aggregate breakage, taking into account interactions among particles and between particles and the surrounding fluid. The steady-state aggregate size distribution closely matches the experimental data (Figure 1 [1, 3]), thereby validating CFD-DEM simulations. The evolving aggregate size and structural characteristics are quantified as they undergo concurrent aggregation and breakage within a linear velocity gradient found in a Couette flow cell, ranging from individual spherical particles to large aggregates, showing strong agreement with experimental observations. The temporal evolution of the aggregate number concentration aligns well with first-order aggregation theory [4] at low shear rates. However, increasing shear rates, increase the deviation from the theoretical collision kernel, due to the reduced collision efficiency at these conditions. Particle dynamics are characterized by three distinct regimes based on shear rate and aggregate size: pure breakage, breakage and restructuring, and restructuring and aggregation (Figure 2 [3]). A novel breakage rate equation is proposed, capturing the transition from the breakage and restructuring regime to the pure breakage limit [3].

References

1. Kushimoto, K., Ishihara, S., Pinches, S., Sesso, M.L., Usher, S.P., Franks, G.V., and Kano, J., *Development of a method for determining the maximum van der Waals force to analyze dispersion and aggregation of particles in a suspension.* Advanced Powder Technology, 2020. **31**(6): p. 2267-2275. https://doi.org/10.1016/j.apt.2020.03.021.

2. Franks, G.V. and Zhou, Y., *Relationship between aggregate and sediment bed properties: Influence of inter-particle adhesion.* Advanced Powder Technology, 2010. **21**(4): p. 362-373. https://doi.org/10.1016/j.apt.2010.02.007.

3. Zeng, L., Franks, G.V., and Goudeli, E., *Aggregation and breakage dynamics of alumina particles under shear by coupled Computational Fluid Dynamics – Discrete Element Method.* Journal of Colloid and Interface Science, 2024. **661**: p. 750-760. https://doi.org/10.1016/j.jcis.2024.01.210.

4. Swift, D.L. and Friedlander, S.K., *The coagulation of hydrosols by brownian motion and laminar shear flow.* Journal of Colloid Science, 1964. **19**(7): p. 621-647. https://doi.org/10.1016/0095-8522(64)90085-6.

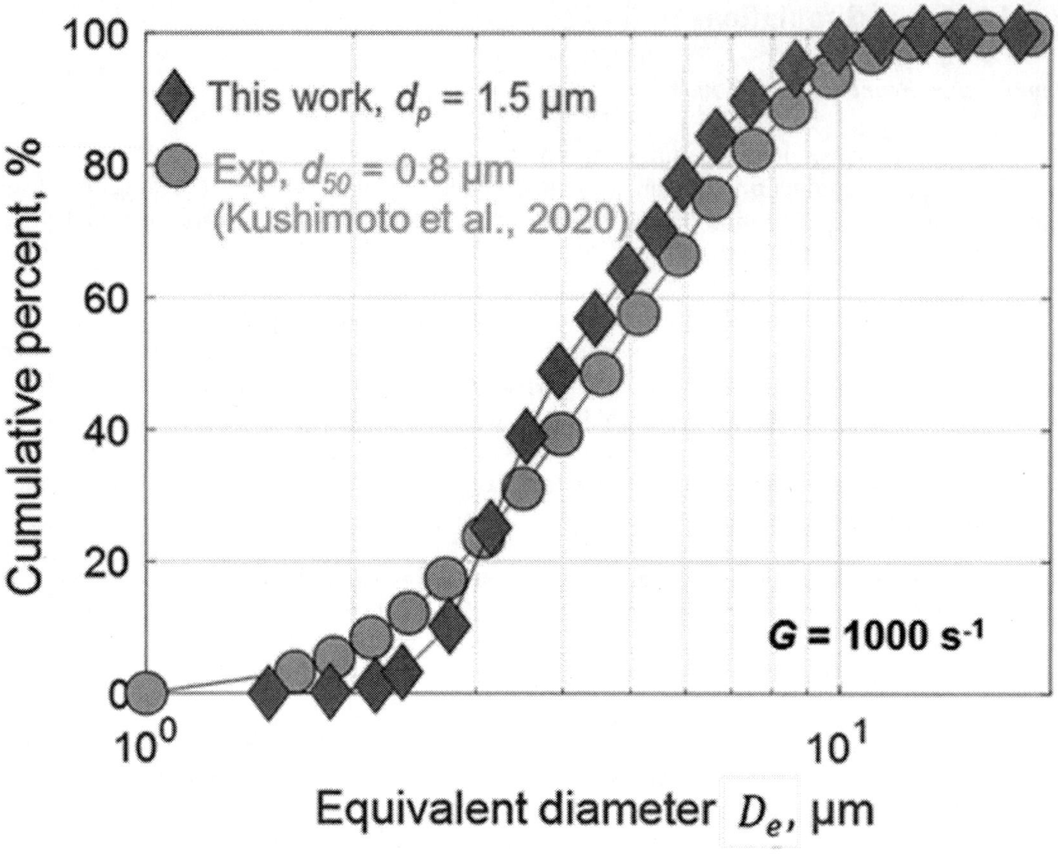

Figure 1. CFDEM-obtained [3] and measured [1] size distribution.

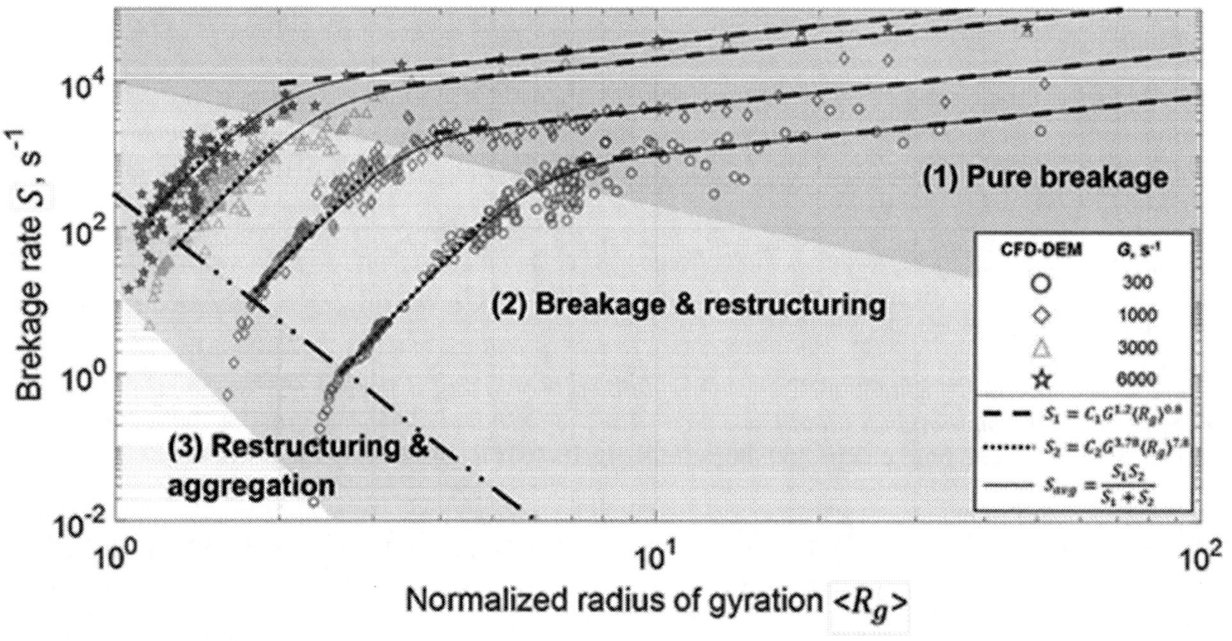

Figure 2. Breakage rate vs. aggregate size [3].

37f Impact of Silica Nanoparticles on Microstructure and Rheology of Wax Network Suspensions

Pooja Saxena, Jyoti Seth, Vinay Juvekar

Suspensions with jammed networks are found in different products such as chocolate, peanut butter, or margarine in the food industries, crude oil-carrying pipelines in the petroleum industries, and creams and ointments in the pharma and consumer goods industries. Organic molecules present in these suspensions serve as crucial structuring agents. The network formed during the cooling plays a pivotal role in determining the texture and rheology of the end products. Common additives like polymers and surfactants are frequently employed to refine the crystal morphology and enhance product properties. Recently, additives containing nanoparticles with different surface characteristics are being explored to improve rheology by modifying the crystal properties.

In the present work, we employ a networked waxy oil system to study the effect of hydrophobic silica nanoparticles (SNPs) on the crystallization behavior of waxes. SNPs have been synthesized using the Stöber method and hydrophobized via attachment of long alkyl chains. We study the influence of alkyl chain length and grafting density on wax crystal morphology and suspension rheology. With the introduction of SNPs in the crystallizing medium, the crystal appearance temperature is reduced, signifying delayed nucleation. Moreover, the morphology of wax crystals changes from plate-like to branched structures. The change in crystal morphology was found to reduce the network-forming tendency of the system with more than one order decrease in viscoelastic moduli and yield stresses. The changes in rheological properties are further correlated to microscopic properties, such as crystal number density, average crystal length, and total free pore volume. SNPs are expected to interfere with the formation of stable nuclei and further adsorb on the growing interfaces of the crystals to inhibit growth in that direction. These results have been confirmed by SEM micrographs obtained for the crystals before and after the incorporation of SNPs. Adsorption on the crystal surface is directly related to the diffusivity and mobility of nanoparticles. Our study also explores the effect of various parameters such as SNP concentration, surface grafting density, and particle size to optimize the efficiency. By shedding light on the mechanisms underlying nano-additive action on wax crystallization, our findings offer valuable insights for the development of tailored solutions to enhance the product performance of similar systems.

548a Shell Thomas Baron Award In Fluid-Particle Systems: Magnetic nanoparticle suspensions in time varying magnetic fields – colloidal hydrodynamics to biomedical applications

Carlos Rinaldi-Ramos

Magnetic nanoparticles respond to time varying magnetic fields via a combination of internal dipole and whole-particle rotation, depending on factors such as thermal motion, hydrodynamic drag, magnetic torques, and internal

barriers to dipole rotation. Depending on the amplitude and frequency of the time varying magnetic field, the nano-particle's response can give rise to the conversion of magnetic field energy into heat or to a signal which can be used

to monitor nanoparticle rotational diffusion or quantify nanoparticle distribution in a subject. This multitude of magnetic responses, coupled with their biocompatibility, make iron oxide nanoparticles of great interest in sensing,

biomedical imaging, drug delivery, and thermal therapy. In this talk I will provide an overview of the theory of magnetic nanoparticle response to time varying magnetic fields and my group's work advancing several of these

applications, including monitoring nanoparticle mobility in complex and biological fluids, cancer thermal therapy, rewarming of cryopreserved organs, and non-invasive, unambiguous, and quantitative tracking of nanoparticles and cells using magnetic particle imaging.

548b PSRI Fluidization and Fluid-Particle Systems Award: Dynamics of fluid-particle systems of magnetically responsive Janus particles

James Gilchrist

This talk will highlight two recent areas of our research on magnetically-responsive Janus particle systems. Fabricated through continuous roll-to-roll deposition and subsequent physical vapor deposition, these silica or polymeric micro- spheres are half coated in iron oxide. At the colloidal scale, suspensions of magneto-responsive Janus colloids form chains and undergo alignment under the influence of a magnetic field. When the magnetic field is aligned with a light path, light transmission through the sample increases as compared to randomly or orthogonally oriented chains. This study investigates the emissivity response of this suspension using visible and infrared light as a function of particle concentration and magnetic field strength. At the granular scale, a magnetic force can be used to tune both the torque and interparticle attraction to tune the motion and friction of dense granular systems. These responsive microroller robotic Janus particles flow with quintessential granular behavior, such as surface layer flow, but against gravity, giving rise to a negative angle of repose and a negative coefficient of friction. These systems also readily segregate and demonstrate creep as granular materials, suggesting granular physics can be used to optimize how to manipulate microroller systems over obstacles and through complex geometries. Rather than studying how robots can navigate granular media, the granular media itself is the responsive material that is manipulated as an amorphous particle- based soft-robot.

548c Elsevier PTF Lifetime Achievement Award

Hamid Arastoopour

A pathway to a sustainable net-zero economy is both multi-faceted and complex, and requires abundant supplies of renewable energy and water. An effective pathway aims to improve the reliability, security, and affordability of energy and water by utilizing a least-cost strategy to reduce the negative impact of energy consumption on climate change and water availability; emphasizing the gradual decarbonization of the global energy system; increasing energy efficiency and conservation and preserving natural resources and the environment. This presentation will discuss examples of the critical role of particle technology research in the creation of a pathway to a sustainable society. Specifically, the presentation will focus on my research group contributions in the areas of computational fluid dynamics (CFD) simulation of CO2 capture in circulating fluidized bed (CFB) systems; CO2 conversion to useful products using an electrolyzer; CFD research in the design of thermal energy storage for concentrated solar energy (CSE) using packed and fluidized bed systems; numerical simulation of the production of natural gas from unconsolidated hydrate reservoirs and particle pulverization as a recycling tool for polymeric and elastomeric materials.

217b Effect of Processing on Physicochemical Properties and Powder Flow Performance of Zeolites

Angeliki Chalasti, Canan Gücüyener, Jerry Heng

Zeolites are hydrated aluminosilicates composed of a crystalline structure of joined silicon oxide (SiO_4) and aluminium oxide (AlO_4) tetrahedra in a three-dimensional network. Thanks to their unique structure of regularly spaced cages and channels of precise sizes and shapes, zeolites can absorb and interact with various elements based on their size, shape, and chemical characteristics. Their high surface area, acidity and excellent thermal stability make them indispensable materials in a wide range of applications in adsorption, ion exchange, and catalysis [1]. Zeolites are produced in batches, and the manufacturing process involves several steps (Figure 1). The zeolite powder, obtained after drying, undergoes high-temperature calcination to create the necessary porosity for its various applications. This process causes a significant change in the physical and chemical properties of the zeolites, which ultimately affects the flow behaviour of the powder [2].

Despite the availability of many techniques for characterizing the physical and chemical properties of zeolites, the comprehensive understanding needed for optimizing manufacturing design and resource utilization during scale-up is limited by the absence of a feedback loop between single particle properties and bulk properties. Although research primarily applicable to pharmaceutical powders has examined how processing affects particle properties like surface (roughness, surface energy, surface area), size and shape [3-6], there remains a gap in similar investigations applied to zeolites, posing significant challenges in predicting the feasibility of utilizing small-scale prepared zeolites on a large scale. In this study, by using zeolites with specific properties, we aim to develop a linkage of various micro-/single properties of zeolites to their bulk powder behaviour under thermal processing (drying and calcination).

Two commercial and eight Johnson Matthey synthesized ZSM-5 zeolites; four after drying and four after calcination (Figure 1), were studied for their physical and chemical properties, including surface energy, surface area, surface chemistry, particle size, density, morphology, and crystal structure. A range of different characterisation methods was employed, such as X-ray techniques (XRD, XPS, XRF), N_2 adsorption, helium pycnometry, imaging techniques (SEM) and particle size analysis to examine the samples before (as made) and after the calcination (calcined) stage. The dispersive surface energy of the samples was determined using inverse gas chromatography (IGC) through the injection of a sequence of non-polar alkane probes (C_5-C_8). Additionally, flow measurements of the zeolite powders were conducted to understand their flow behaviour under dynamic and shear conditions. A stability and variable flowrate test was performed using an FT4 powder rheometer to determine the basic flowability energy (BFE) of the samples at varying tip speeds during the downward movement of the blade through the powder bed. For the shear testing, an Anton Paar shear cell was used to determine the flow function coefficient (ff_c) of the samples composed of 5 points (1, 2, 3, 4 and 5 kPa). Shear steps were taken at three applied normal stresses, at 30%, 50%, and 70% of the preshear normal stress, to determine every yield locus.

Regardless of the preparation history of the samples, the process of calcination resulted in a significant increase in their surface area and porosity. The average surface area of as-made samples was 54 m^2/g, which increased to 410 m^2/g after calcination. Similarly, porosity increased by~20% with this process. This was in line with the expected result of the organic template being removed from the zeolite structure. The surface energy of the samples also increased after calcination by an average of~98 mJ/m^2. The bulk and surface properties of zeolites 79A and 79B appeared to be significantly influenced by the source of aluminium used during production. For sample 79A, $Al(NO_3)_3 \cdot 9H_2O$ was used as the Al source, whereas sample 79B was synthesised with $Al(OH)_3$. Both materials differed in terms of particle size distribution, surface area, and surface energy yet exhibited the same gel SAR (~90) and percent crystallinity (100%) (Figure 2). N_2 adsorption analysis of the zeolites before and after calcination showed that 79B had a greater external surface area, surpassing 79A by a margin of 143% pre-calcination and 26% post-calcination (Figure 2), while maintaining a closely similar microporous area. This suggests that agglomeration evolved during calcination, which was more prominent for zeolite 79B than 79A. This was accompanied by a sharper increase in its surface energy post-calcination increasing by 149% in contrast to 15% for sample 79A.

Powder flow measurements of the materials indicated that samples T840 and 82A as made exhibited the best flow performance with average BFE values of 43 mJ and 60 mJ, respectively. This was attributable to their larger particle sizes (D50 ~10µm) in comparison to the other samples. 79A and 79B as made demonstrated the poorest flow (BFEs of 264 mJ and 330 mJ), which is in line with their small particle size (D50 ~1µm). Examining the powder flow behaviour using shear testing suggested comparable trends (Figure 3). Except for T840 and 82A which exhibited a free flowing flow (ff_c~10), most samples fall into the very cohesive region ($1<ff_c<2$) (Figure 3). Calcination had a significant effect on flow behaviour, leading to decreased BFEs, or in other words an increase in the ff_c and enhanced powder flow. The initial findings indicate a correlation (r=-0.8267, p=0.0114) between the enhancement in powder flowability and the increase in zeolite porosity through calcination. This was further supported by the decrease in bulk density observed after calcination, which positively correlated with the BFE (r=0.8447, p=0.0083), leading to improved flow. Additional experiments are underway to validate the observed differences by exploring binary zeolite mixtures under high-temperature conditions.

In summary, this work highlights the effect of particle handling methods and processing, mainly calcination, on zeolite properties during manufacturing. Analysis of zeolites with different processing histories and forms showed that surface energy, surface area, and porosity increased after calcination. Zeolites 79A & 79B prepared with different Al sources exhibited differences in surface properties. Particle properties exerted an impact on the flow behaviour under various conditions, emphasizing the critical role of physicochemical characteristics in determining or even predicting powder performance at an industrial scale.

References

[1] E. Pérez-Botella, S. Valencia, F. Rey, Zeolites in adsorption processes: State of the art and future prospects, Chemical Reviews 122 (2022) 17647–17695.

[2] S. Zinatlou Ajabshir, C. Gucuyener, V. Vivacqua, D. Gobby, H. Stitt, D. Barletta, M. Poletto, Flow behaviour of zeolite powders at high process temperatures, Powder Technology, 409 (2022) 117818.

[3] U. V. Shah, V. Karde, C. Ghoroi, J. Y.Y. Heng, Influence of particle properties on powder bulk behaviour and processability, International Journal of Pharmaceutics, 518(1–2) (2017) 138–154.

[4] H. Salehi, V. Karde, H. Hajmohammadi, S. Dissanayake, S. H Larsson, J. Y.Y. Heng, M. Bradley, Understanding flow properties of mannitol powder at a range of temperature and humidity, International Journal of Pharmaceutics, 596 (2021) 120244.

[5] U. V. Shah, D. Olusanmi, A. S. Narang, M. A. Hussain, M. J. Tobyn, J. Y.Y. Heng, Decoupling the contribution of dispersive and acid-base components of surface energy on the cohesion of pharmaceutical powders, International Journal of Pharmaceutics 475(1–2) (2014) 592–596.

[6] V. Karde, A. E. Jefferson, G. A. Hebbink, J. Y.Y. Heng, Investigating sizing induced surface alterations in crystalline powders using surface energy heterogeneity determination, Powder Technology, 395 (2022) 645-651.

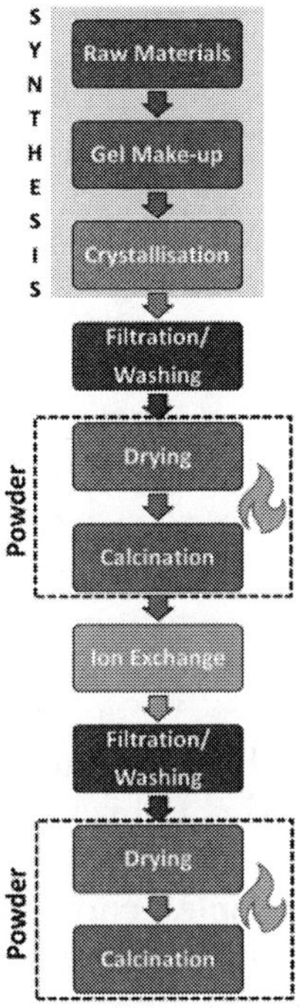

Figure 1: Schematic of the multi-stage process during zeolite manufacturing, highlighting the thermal processing steps involved.

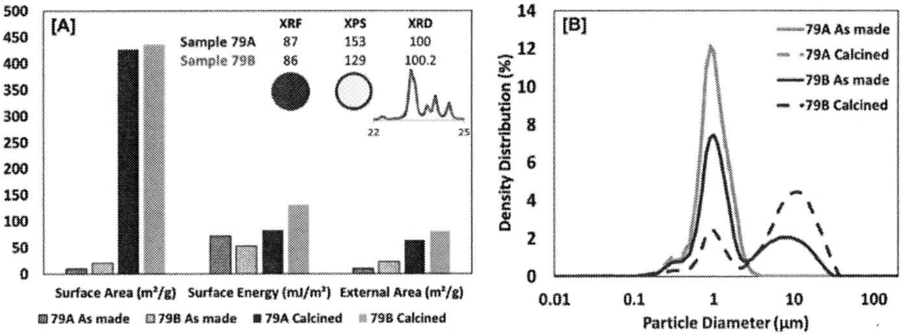

Figure 2: *[A] Surface area, surface energy and external area along with bulk Si:Al ratio (XRF), surface Si:Al ratio (XPS) and relative crystallinity (XRD) data,* ***[B]*** *particle size distribution for samples 79A and 79B before and after calcination.*

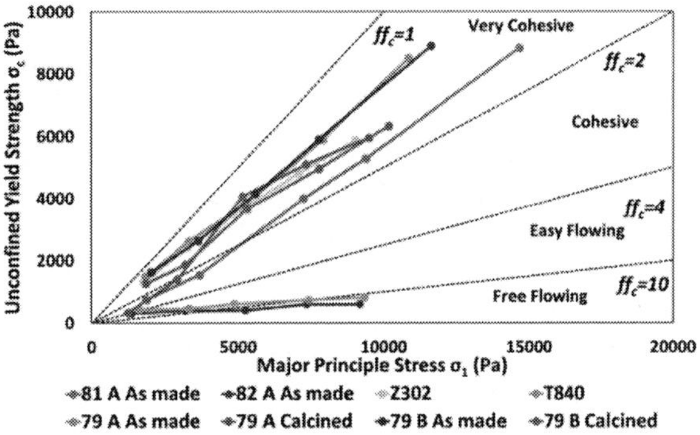

Figure 3: *Flow functions of ZSM-5 zeolites using the Anton Paar shear cell.*

217c Food Phantom Packaging: Prolonging Product Shelf Life Via Microstructure Engineering of Particulate Consumer Foods

Luc Dewulf, Michael K. Hausmann, Annabel Bozon, Gerhard Niederreiter, Agba D. Salman

Driven by consumer trends and environmental commitments to reduce conventional plastic and metal packaging in the food supply chain, consumer food companies are racing to develop alternative packaging solutions, mainly focussing on paper-based substitutes [1].

A paradigm shift in packaging innovation integrates the barrier properties of packaging into the food microstructure, fulfilling an invisible protection from the environment while reducing dependence on external conventional wrapping material, thus enabling increased packaging paperisation or even zero-waste schemes. We demonstrate the successful application of our microstructure engineering approach for oil containing seasoning tablets and show that unwanted oil-release into absorbing paper packaging can be reduced and unpleasant oil stains entirely prevented. For example, by reducing the food particle size of crystalline ingredients tenfold from $D50 = 500$ to 50 μm, we managed to achieve a 100% retention of oil within the food structure, thereby protecting the packaging from negative visual effects. In accelerated shelf-life tests the oil-release into sample paper wrappers was monitored via first-of-its kind Raman Laser Chemical Imaging coupled with automated optical image analysis (Fig. 1 a, b). The capillary flow phenomenon of oil in the food structure was modelled fluid-mechanically from first principles, correlating microstructural properties with shelf-life performance (Fig. 1 c). While a multitude of prior research has been conducted on the process of porous media imbibition with liquids in the area of engineering materials [2] or food [3], the phenomena of liquid release from food microstructures into a second porous medium such as packaging is little reported.

As such, this work contributes to explain food metastability resulting from liquid transport through model food tablets with the aim to harness particle engineering to innovate more shelf-stable food microsystems for a waste-free packaging future, ensuring food safety, affordability, and consumer acceptance.

[1] B.L. Tardy, *et al.*, *Nature Sustainability*, 6 (2023) 360-367.

[2] S. Gruener, P. Huber, *Transport in Porous Media*, 126 (2019) 599-614.

[3] J. Kammerhofer *et al.*, *Powder Technology*, 339 (2018) 765-774.

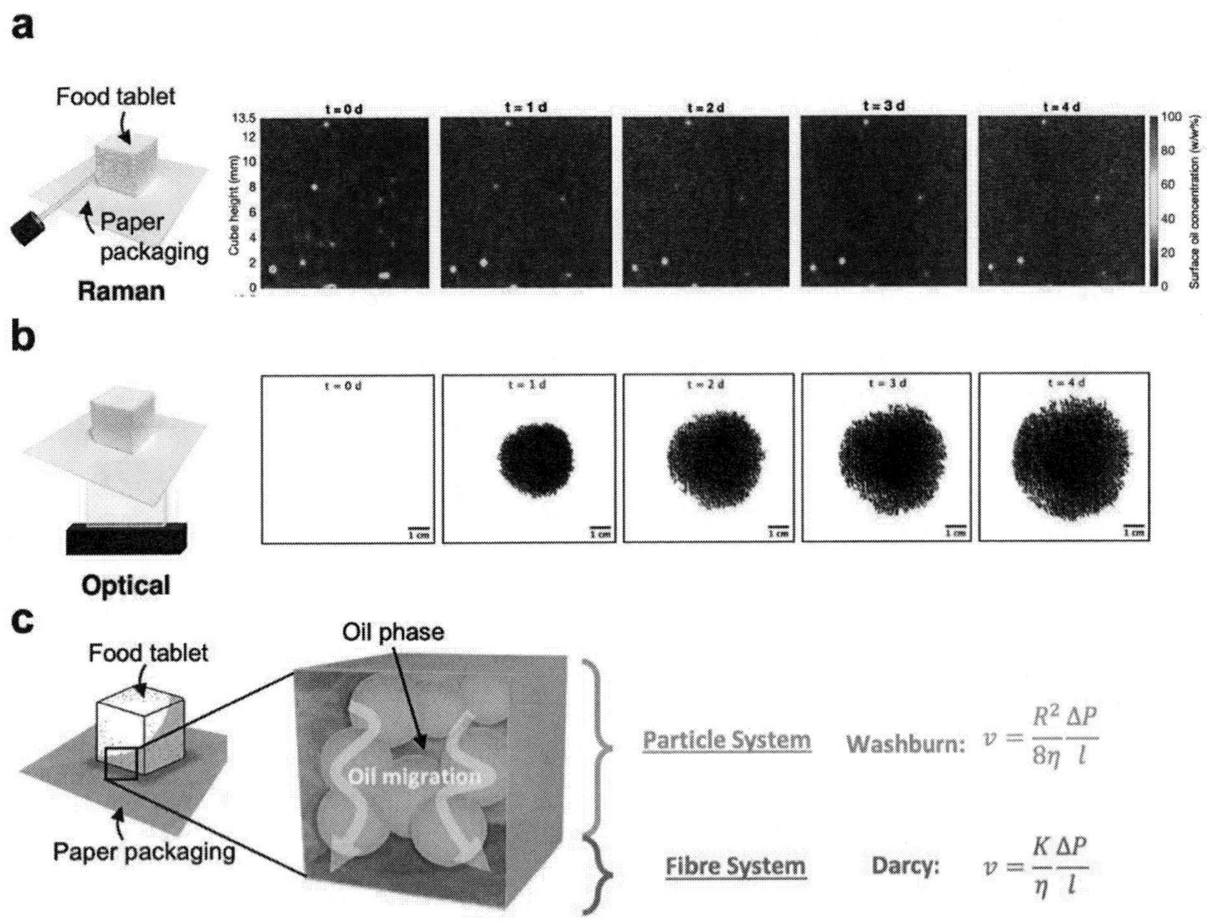
a
Food tablet
Paper packaging
Raman
t = 0 d
t = 1 d
t = 2 d
t = 3 d
t = 4 d
Cube height (mm)
Surface oil concentration (w/w%)
b
Optical
1 cm
c
Food tablet
Oil phase
Oil migration
Paper packaging
Particle System
Washburn: $v = \frac{R^2}{8\eta}\frac{\Delta P}{l}$
Fibre System
Darcy: $v = \frac{K}{\eta}\frac{\Delta P}{l}$

217d The Effect of Moisture Content on Mechanical Properties of Alumina Catalysts Tablets

Mingzhi Zhao, Sai Yellapragada, Christine Hrenya, Benjamin Glasser, Alberto Cuitino

Aluminum oxide, widely known as 'Alumina,' has extensive use in the chemical and petrochemical industries. Within these sectors, compacted aluminum oxide tablets are commonly used as catalyst support to facilitate efficient chemical processes. These tablets are engineered to meet specified mechanical properties, including hardness, tensile strength, Young's modulus, and porosity. Manufacturers often need to adjust the compaction load, amount of binder and lubrication to optimize the manufacturing process to produce alumina tablets with optimal mechanical characteristics, ensuring enhanced performance and durability while keeping the cost relatively low.

This work investigates the effect of moisture content on various type of alumina tablets. The moisture content of alumina powders can change due to humidity as well storage and processing conditions. The tablets underwent compaction with differing moisture contents and compaction loads and were subsequently tested 24 hours post-ejection. Elevating the moisture content up to 20% w/w resulted in an increase in both the tensile strength and Young's modulus of the tablets.

Our hypothesis suggest that the additional liquid acted as a lubricant allowing the particles to move past one another and gas to escape and thus form a denser packing structure. Moreover, drying tablets at 120°C for 24 hours effectively removes most of the water, thereby altering their mechanical properties. Notably, to the best of the authors' knowledge, there has been a dearth of studies focusing on the effect of moisture on alumina tablets. Our results show that controlling the moisture content is important in order to produce tablets with low variability. In addition, our results show that increasing the moisture content can be used as a strategy to increase the strength of the tablets.

217e Atomic Layer Deposition of Sintering Aids to Accelerate Tungsten Powder Metallurgy

Davis R. Conklin, Hailey Loehde-Woolard, Arne Croell, Jhonathan Rosales, Alan W. Weimer

Tungsten is a dense, highly refractory metal used in a wide range of engineering applications where high temperatures are present. Due to its high melting point, high hardness, and poor ductility under normal conditions, there is interest in producing near-net shape tungsten parts using powder metallurgy to reduce or eliminate difficult machining operations. During this process, tungsten powder is pressed into a desirable shape and sintered at high temperature to bond the particles and remove porosity. Trace amounts of transition metal sintering aids such as Pd or Ni can greatly accelerate this process by forming a thin intergranular film that provides a fast diffusion pathway for tungsten atoms below the eutectic point. This work deposits small amounts of Pd onto tungsten powder using particle atomic layer deposition (ALD), which ensures uniform coverage of the primary tungsten particles and minimal effects on power properties relevant for downstream powder metallurgy processing. The morphology of the ALD Pd coating is shown to enhance the performance of the sintering aid by lowering the temperature and which sintering occurs.

This work is the first use of particle ALD to coat a dense, cohesive refractory metal powder. Pd was deposited using alternating doses of $Pd(hfac)_2$ and formalin vapors through a heated fluidized bed of 1-5 μm tungsten powder. A composition of approximately 0.1wt% Pd was achieved with 25 ALD cycles. For comparison, a sample with matching composition was prepared by immersing the W powder in a Pd salt solution and drying. The powder compositions were characterized by ICP-OES and LECO oxygen analysis, and the morphology will be characterized by surface area analysis, SEM, and TEM. The sintering kinetics of the powders were characterized using dilatometry up to 2000°C. Analysis of the sintering rate profiles suggests that the lower-temperature sintering of the ALD Pd-W powder is not due to a change in mechanism. Instead, the morphology of the ALD Pd coating likely assists the formation of the Pd-rich intergranular film responsible for accelerated tungsten diffusion. This is supported by further analysis of fully sintered samples using EDS, which reveals a more uniform distribution of Pd in the final tungsten microstructure when using the ALD Pd-W powder.

These findings will help improve near-net shape tungsten manufacturing by lowering the required sintering temperatures and times to achieve a dense part. Lowering the cost of tungsten parts will allow more of their unique properties to be fully utilized by the engineering community and accelerate development of high-temperature components.

217f Processing and Design Challenges of Highly Filled Composites

Hulusi Turgut, Kelly Krzysik, Bryan Kenyon

Highly filled composites are densely packed material systems that approach the maximum achievable packing fraction of the solid within the composites. These material systems have found significant applications in aerospace, automotive, construction, cosmetics, marine, electronics and biomedical. Examples of applications that utilize highly filled composites to engineer the next generation of materials include prosthetic, dental, wood-substitute, conductive, coating, and structural composites. The processing of highly filled composites presents several challenges due to the high viscosity and poor interaction between polymer matrix and solid particles. Some of the processing challenges include the inability to achieve a uniformly filled polymer matrix, an acceptable filtration rate, reliable quality measurements, and long-term stability without phase or filler separation. These challenges can be addressed by utilizing a fundamental understanding of material science and processing technologies. Herein, we report some of the key processing and material science challenges for the commercialization and manufacture of highly filled composite systems.

341a Mixed Soft-Rigid Granular Flows: Role of Soft Particles in Inducing Metastable Arches

Saeed Alborzi, Sara Hashmi

Mixing soft and rigid in the flow of granular particles in a 2D hopper reveals interesting, counter-intuitive features about the role of each particle type in flow and clogging; in stable arches that cause permanent clogs, the arch composition i.e. the probability of soft or rigid particles composing a stable arch is equal to the composition of the original mixture [1]. Also, arch shapes vary more in rigid particles than the soft, giving higher chances to the rigid to form a clog [2]. However, not all arches are stable when introducing soft particles to the ensemble. Metastable arches occur when a set of particles meet near the exit and collapse spontaneously with some delay. We characterize this phenomenon in the discharge of mixed cm-sized soft and rigid particles in a top-down symmetrical 2D hopper.

We find that perfectly rigid ensembles flow faster but clog more, while soft particles drain more slowly and intermittently with less clogging: a well-known notion in granular flows saying "Faster is slower!". The distribution of time intervals t between the exiting particles exhibits a power-law behavior regardless of the mixture composition. Although soft particles are less stable in permanent arches, we show that they are equally stable as rigid particles in metastable arches, rather the lowest angle between the arch particles determines which particle is stable the least.

[1] Alborzi, S., Clark, B.G. and Hashmi, S.M., 2022. Soft particles facilitate flow of rigid particles in a 2D hopper. *Soft Matter*, *18*(21), pp.4127-4135.

[2] Alborzi, S., Abrahamyan, D. and Hashmi, S.M., 2023. Mixing particle softness in a two-dimensional hopper: Particle rigidity and friction enable variable arch geometry to cause clogging. *Physical Review E*, *107*(2), p.024901.

341b Segregation of Fine and Cohesive Particles

Richard Lueptow, Qiong Zhang, Dhairya Rajendra Vyas, Paul Umbanhowar, Julio M. Ottino

Great strides have been made in understanding and predicting size-segregation in flows of non-cohesive particles with size ratios $R < 3$. Extending these advances to cohesive particles and larger size ratios (fine particles) is difficult because of experimental and simulation challenges as well as the wide range of possible particle properties and flow conditions. Here we report preliminary results from DEM simulations for size-bidisperse cohesive particles and for non-cohesive fine particles with size ratios as large as $R = 40$. The usual segregation pattern in heap flow of small particles depositing on the upstream portion of a heap and large particles depositing further downstream occurs for low cohesion between particles for $R = 2$. When the cohesion is increased while maintaining the same feed rate, alternating layers of large and small particles result due to intermittent flow down the heap. Higher levels of cohesion prevent segregation. In general, the flow and segregation behavior are determined by the ratios of the cohesive force to the particle weight and to the flow-related shear force. Furthermore, the relative strengths of large-large, small-large, and small-small particle cohesive interactions must be considered. For large size ratios without cohesion, the segregation or lack thereof, in a static large particle bed under gravity depends on a complex combination of size ratio, flow conditions, particularly the magnitude of the small particle velocity fluctuations, and particle properties, including the friction and restitution coefficients. Nevertheless, we have been able to determine scalings that account for these various conditions. This material is partially based upon work supported by the National Science Foundation under Grant CBET-2203703.

341c Scaling Continuous Powder Blenders: The Length-Scale Problem

Ivan Bogaerts, James Holman, Anthony Tantuccio, Sarang Oka

With growing adoption of pharmaceutical continuous manufacturing, understanding transferability of processes between manufacturing lines is becoming of critical importance. This comprehension facilitates tech transfer of processes between lines of similar or dissimilar scales hosted at different sites or to or from a CDMO. Development of robust, science-based approaches to facilitate the transfer of continuous processes is key to growing adoption of the technology.

Powder blending is a critical unit operation in pharmaceutical manufacturing of oral dose products. Blending in continuous processes is most commonly performed in tubular blenders. These blenders, fitted with an impeller facilitate mixing along two length-scales – radial and axial. Substantial progress has been made in understanding how powder mixing scales on the radial axis, that is how mixing scales in blenders of increasing diameters. However, little is understood on how mixing scales along length of the blender. This work attempts to understand mixing progression along the axial direction of a continuous powder blender.

This examination is performed by leveraging experiments on two inclined continuous powder blenders of different lengths but the same diameters. Two powder throughputs are explored in combination with three different blade configurations. Blending performance is measured by virtue of measuring the mass holdup, mean residence time and the number of blade passes. Early work on understanding axial dispersion along the length is also discussed.

341d Investigation of Mixing Behavior in Shape-Dependent Polydisperse Mixtures and Critical Length Scale of Percolation in a Vibrating Bed System Using DEM

Praveen Dubey, Anshu Anand

The present study employs DEM simulation to explore the interplay between shape-dependent size polydispersity, the critical length scale of percolation, and mixing behavior within a vibrating packed bed system. In the realm of mixture types, the investigation differentiates between homogeneous (single shape) and heterogeneous (multiple shapes) mixtures, providing insights into their distinct effects on mixing. Non-spherical particles are generated using Super-quadric methods. Size polydispersity is represented by the standard deviation to mean particle diameter ratio, and the impact of particle shape on mixing are assessed using the Overall Mixing Index (SMI). Findings reveal a decrease in mixing as polydispersity transitions from a narrow to a wide distribution in homogeneous Gaussian mixtures. The investigation delves into the effects of coarse and fine particle shapes—sphere, ellipsoid, cylinder, and cube—on mixing behavior for heterogenous gaussian mixtures. Another focal point of the research is the critical length scale of percolation, representing the minimum size of the dynamic void allowing fine particles to percolate freely in a series of binary mixtures. The study scrutinizes the impact of key particle parameters (shape and size) on this critical length scale. For coarse particle shapes, certain shapes appear to influence the critical length scale differently than others, and the introduction of intermediate-size particles decreases the critical length scale of percolation. A predictive machine learning model has been developed to directly estimate the percentage of percolated fines using input variables such as size ratio and sphericity ratio. By analyzing the percentage of percolated fine particles, the critical length scale of percolation is determined for various combinations of size and sphericity ratios. The results indicate that employing nonspherical particles in the vibrated packed bed system can minimize the critical length scale of percolation. This comprehensive investigation provides insights into the intricate dynamics governing mixing behavior, emphasizing the role of particle size distribution, shape, and percolation phenomena in vibrated packed bed systems.

341e Exploration of the Effect of Micro Dimpling on Wall Materials to Improve Flow of Cohesive and Non-Cohesive Powders in Bins and Hoppers

Kevin Kellogg, Justin Phillips, Matthew Dubay, Joerg Heinrich, Jorg Theuerkauf

Micro dimpling can be used to create surfaces with reduced effective friction for solids flows. Therefore, micro dimpling is a candidate for surface treatment of the walls of bins and hoppers to improve solids flow without using a coating. In the current work, we experimentally evaluate the effective wall friction reduction associated with micro dimpling using Schulze Ring Shear testing and a Jenike Shear testing to measure the wall friction angle. Specifically, we measure the wall friction angle of bulk solids on wall coupons before and after micro dimpling. To evaluate whether the reduction in effective wall friction angle is due to a reduction in surface friction or a reduction in adhesion, the wall friction angle is measured for both cohesive (limestone) and non-cohesive (silica sand) bulk solids. We also consider different initial finishes (prior to micro dimpling) of the wall coupons. We find that micro dimpling does not improve the effective wall friction angle for certain initial coupon finishes.

234b Are Blenders Needed in Continuous Direct Compaction?

Yi Tao, Carlos Ortega-Zuniga, James Scicolone, Fernando Muzzio

Since conception, continuous pharmaceutical manufacturing has remained in the shadow of batch manufacturing. Traditional feasibility studies, validation runs, and regulation oversight have contributed to a delay in widespread adoption. One such area of old-fashioned and usually fixed thinking is blending. In batch manufacturing, ingredients are added individually, in large masses, into bins, requiring significant blending to achieve a homogeneous state. On the other hand, in continuous manufacturing, individual components are usually added at less than 50 grams per second to a continuous process stream that regularly has less than 3 kg of material in transition throughout the manufacturing line at a given time. While blending in batch processes has been well established due to its simplification, the dynamics of continuous manufacturing introduce complexities that challenge traditional thinking. Residence time distribution is one of the critical concepts in continuous manufacturing. Evaluation of residence time distribution using spectroscopic unique tracers has shown that the tracer distribution curves appear as sharp, distinct, peak in concentration with a slight tail exiting the continuous blender. The tracer continues down the CM line, primarily in a plug flow state, until it reaches the feed frame of the tablet press. When the concentration curve of the tracer is evaluated in the feed frame , the peak of the curve has decreased by over two thirds, has a wide distribution, and is commonly a long tail. This concentration curve is similarly observed in the tablets collected after compaction. The evidence identifies that more mixing occurs in the feed frame of the tablet press than in the continuous blender, leading to the question. *Is a blender necessary for a continuous manufacturing line?*

This study the mixing capabilities of a continuous manufacturing line of loss-in-weight (LIW) feeders dispensing into a continuous blender and tablet press, an IBC bin blender dispensed directly into a tablet press, and LIW feeders dispensing individual components to a tablet press with a single pitch blade mixer placed in the hopper of the tablet press were compared to the mixing capabilities of LIW feeders dispensing directly into a tablet press, with only the feed frame of the press to provide the mixing. The formulation utilized for this study was a well flowing, high-dose formulation consisting of 50 %w/w Compap L™, 44 %w/w Prosolv® HD90, 5 %w/w pregelatinized starch, and 1 %w/w MgSt. Blend homogeneity (BH) was monitored within the feed frame of a tablet press and the critical quality attributes of the tablets, such as weight variability, hardness variability, tensile strength, and API content, were evaluated. The results showed that the blend homogeneity for each manufacturing process reached a similar level of homogeneity. The findings indicate that using LIW feeders to dispense directly into the tablet press can generate a homogeneous blend that can match BH for batch and continuous blending. The results of this study provide a potential alternative to significantly reduce manufacturing costs and allow for more rapid product development.

342a Adaptive PBM-DEM Framework for Wet Granulation: A Mechanistic, Multi-Component, and Bi-Directional Approach

Ashok Das, Tarun De, Ashley Dan, Rohit Ramachandran

A novel bi-directional multi-component coupling framework integrating Population Balance Modeling (PBM) and Discrete Element Method (DEM) for simulating wet granulation processes is introduced. In conventional approaches, liquid droplets are treated as granular particles in DEM, resulting in significant computational demands due to liquid droplet generation and deletion processes. To address this challenge, a novel method for incorporating liquid droplets within the PBM is proposed, enhancing the framework's ability to accurately simulate particle-liquid interactions crucial for granulation outcomes. Additionally, an adaptive system scaling method is introduced within DEM, which dynamically adjusts system dimensions to accommodate a fixed number of particles, reducing computational resources by 82% while maintaining simulation accuracy. By integrating these advancements, the effects of key process parameters, such as liquid-to-solid ratio and addition timing, on granulation processes are investigated. This research not only enhances understanding of granulation but also offers potential for advancements in pharmaceutical manufacturing.

342b Extraction of Aggregation Kernel from DEM Simulations Using Sustained Contact: A New Kernel Extraction Methodology for Drum Granulators

Archita Karar, Jayanta Chakraborty, Jitendra Kumar, Anurag Tripathi

Drum granulator is a popular granulation equipment whose modeling is usually conducted using the population balance model. The population balance model requires the knowledge of breakage, aggregation, growth and nucleation kernels. Such kernels are either developed through phenomenological models [1] or inverse problems [2]. Additionally, DEM simulations can be used to model these kernels [3]. While numerous studies exist on deriving aggregation kernels from DEM simulations [4, 5, 6], they predominantly rely on relative velocity, specifically, the collisional velocity between particle pairs, to obtain these kernels. However, these methodologies cannot capture the physics of sustained contacts and resulting liquid/solid bridge formation. Hence a reliable method for kernel extraction is needed for systems where aggregation happens due to sustained contacts. In this work, we demonstrate a new method for the extraction of aggregation kernels using information about sustained contacts.

In the proposed approach, aggregation kernels are obtained as a time snapshot by analyzing the particle locations over a certain number of time steps ($n\Delta t$) around that particular time. The location of the particles provides the duration and average overlap and a few additional information on the contact (such as the number of attachments/detachments). Subsequently, this information is processed through a series of physically grounded criteria to identify contacts likely to result in successful aggregation, thus determining the aggregation efficiency. Finally, the obtained information is translated into the aggregation kernel by using appropriate expressions. This simple contact information-based kernel extraction methodology successfully reproduces several established kernels. To illustrate, we apply this methodology to extract the Brownian kernel (see Figure 1). After validating our kernel extraction method for a few additional known kernels, we extract the aggregation kernel for the case of a rotating drum containing wet particles. We then compare it with the kernels used in literature for granulation in case of dense flows.

References :

[1] Ramkrishna, D. (2000), Population balances: Theory and applications to particulate systems in engineering, Elsevier.

[2] Ramkrishna, D. (2000), Inverse Problems in Population Balances. In Population balances: Theory and applications to particulate systems in engineering (pp. 221-273), Elsevier.

[3] Gantt, J. A., Cameron, I. T., Litster, J. D. & Gatzke, E. P. (2006), 'Determination of coalescence kernels for high-shear granulation using dem simulations', Powder Technology 170(2), 53–63.

[4] Gantt, J.A. and Gatzke E. P. (2006), ' A stochastic technique for multidimensional granulation modeling', AIChE Journal, 52(9):3067 – 3077. doi: 10.1002/aic.10911.

[5] Reinhold, A. & Briesen, H. (2012), 'Numerical behavior of a multiscale aggregation modelâA˘Tcoupling population balances and discrete element models', ˇ Chemical engineering science 70, 165–17

[6] Barrasso, D. & Ramachandran, R. (2015), 'Multi-scale modeling of granulation processes: bi-directional coupling of pbm with dem via collision frequencies', Chemical Engineering Research and Design 93, 304–317

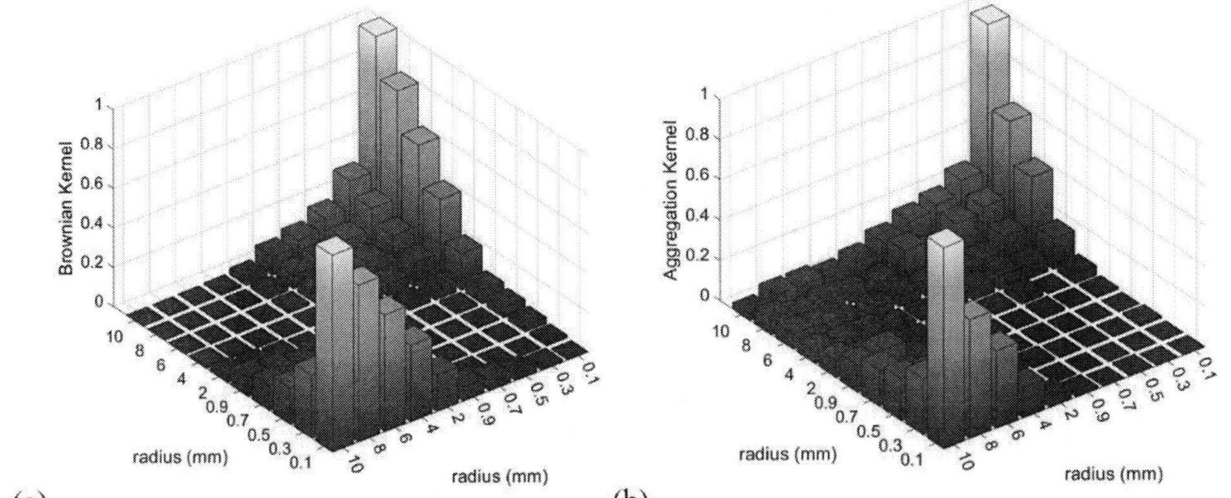

(a) (b)

Figure: 1 : Normalized (a) Brownian kernel and (b) DEM extracted aggregation kernel

342c Combining Artificial Intelligence and Discrete Element Method Simulations for Oral Solid Dose Manufacturing Process Optimization - a Case Study on Bin Blending

Stefan Pantaleev

The rapid and cost-effective design of reliable manufacturing processes is key to maximizing the profitability of pharmaceutical products, a large proportion of which are in the Oral Solid Dose (OSD) form. OSD manufacturing processes pose significant challenges because of the complex nature of the particulate materials they manipulate. The mechanics of these processes is still not fully understood leading to over-reliance on expensive and time-consuming physical trials in the process development stage where delays can be costly due to time limits on patents and competitive market conditions.

Virtual process design and optimization using Digital Twins (DTs) can reduce development timelines and costs and has become an important component of the digital transformation strategy in the pharmaceutical industry. Digital Twins based on high-fidelity Discrete Element Method (DEM) simulations have already been used to reduce manufacturing costs and time to market for new OSD products, but purely DEM simulation-based DTs can be computationally expensive and require expertise to develop and deploy. Going beyond the purely simulation driven approach to generate Artificial Intelligence (AI) based DTs from DEM simulation data can be a much more efficient approach.

In this work we present an efficient virtual optimization methodology for OSD manufacturing processes which combines DEM modelling, design of experiments, AI and optimization methods in the Altair portfolio of tools to significantly reduce the computational expense of optimization relative to a purely simulation driven approach. A key feature of the methodology is the generation of an AI based process Digital Twin, which is a multi-layer perceptron model that constitutes a state-space representation of a dynamic system and is trained on synthetic data from a statistically efficient set of high-fidelity DEM simulations. Because of its high computational efficiency, the AI based DT can be used in conjunction with a gradient descend optimization algorithm to identify the globally optimum operational parameter set in seconds rather than the weeks required by the equivalent purely simulation-based approach.

The validity of the methodology is evaluated through the optimization of an industrial scale bin blending system, which is commonly employed in OSD manufacturing operations. Batch failures are a persistent problem in this type of system, but its physical prototyping is expensive, making it an excellent candidate for virtual optimization. The work focuses on improving the mixing rate in the system by optimizing operational parameters such as the rotational velocity, the level of fill and the bin orientation. The accuracy of the predicted optimal operational parameter set is evaluated in silico, and the advantages and limitations of the methodology are discussed. Finally, ways to effectively deploy and democratize this technology through customized dashboarding are explored.

342d A Model for Reliable Prediction of Tabletability of Mixtures from Those of Individual Components

Changquan Sun

There is a pressing need in the pharmaceutical industry for developing a tablet more quickly while using less active pharmaceutical ingredients (API). The empirical approaches that heavily depend on the use of Design of Experiments (DOE) cannot meet this need. Instead, a materials science-based approach to enable fast and material-sparing tablet formulation development is highly desired. In this context, the ability to predict the compression behaviors of a mixture from that of the individual components and the composition of the mixture is extremely useful.[1] In this presentation, I will describe an approach for successfully predicting tabletability of mixtures consisted of common pharmaceutical excipients and APIs.

Tabletability data was generated using a compaction simulator (Styl'One Evolution, MedelPharm, Beynost, France) simulating a Korsch XL100 compression cycle at 20 rpm with a 10 MPa precompression step. The main compaction pressure varied between 25 and 500 MPa. Round, flat-faced (6 - 11.28 mm in diameter) punches were used for all compaction. External lubrication (magnesium stearate spray, Styl'One MIST) was used on the die wall and punch tips before each compression. Tablets were broken with a texture analyzer, and tensile strength was calculated using the breaking force along with tablet dimensions following a standard process.

The first step of the prediction exercise involves quantifying tabletability data (tensile strength versus main compaction pressure) of individual powders with the Vreeman-Sun equation.[2] Nonlinear regression of tabletability data using the Vreeman-Sun equation leads to three fitted parameters (*P*), $_{max}$, , and . to completely describe tabletability. This set of tabletability parameters of any two powders were used to predict a set of parameters of a mixture between the two powders using the power mixing law based on composition (*C*), i.e., $P_{1,2}=P_1^{C1}P_2^{C2}$.

The tabletability of a large number of binary mixtures for several pairs of powders exhibiting a wide range of mechanical properties and tabletability were successfully predicted. Validity of this approach was also shown for ternary and five-component systems, suggesting robustness and general applicability of this modeling approach.

Acknowledgments

This work was supported by the Center for Integrated Materials Science and Engineering for Pharmaceutical Products (CIMSEPP), which is funded by the National Science Foundation (grant IIP-2137264). Contributions from Gerrit Vreeman, Vedant Bhagali, Aakash Hasabnis, and Zijian Wang are greatly appreciated.

References

1. Ramaswamy, C. M., Varma, Y. B. G. & Venkateswarlu, D. Compaction of mixtures of materials. *Chem. Eng. J.* **1**, 168–171 (1970).
2. Vreeman, G. & Sun, C. C. A powder tabletability equation. *Powder Technol.* **408**, 117709 (2022).

342e A Reduced-Order Mechanistic Tablet Coating Model to Capture Downstream Drug-Release Variability.

Abhishek Paul, Indu Muthancheri, Mario Rousselin, Yasser Jangjou

The drug release process of pharmaceutical tablets is divided into three phases: the lag phase, the zero-order phase, and the decaying phase. In the case of coated tablets, the underlying mechanism of release can be governed by osmotic pumping or diffusion, depending on the permeability of the coating layer. Although mechanistic models to predict drug release profiles of film coated tablets have been proposed, the expected variability in release profiles for a batch of coated tablets is an open question [1]. Furthermore, understanding the impact of tablet coating on the expected variability in release profiles is important for better process control on the coating process, especially at different operating scales. A model-driven tablet coating approach will aid in achieving a tightly controlled drug release profile for the entire batch, especially in the case of controlled-release tablets. In order to develop a quality by design (QbD) approach for tablet coating and dissolution, it is necessary to examine the variability of both processes.

First, we adopt a reduced-order mechanistic model to describe the quality of the tablet coating process in terms of dimensionless numbers [2]. Second, we measure the drug release profile of a randomly selected sample set of the coated tablets. We then statistically investigate the correlation between the process parameters, the coating dimensionless numbers, and the measured variability in the drug-release profile of the batch. From the results, we examine whether inter-tablet coating variability (i.e., the nonuniform coating layer thickness across tablets) indicates downstream dissolution variability. In this context, inter-tablet variability is represented by the coefficient of variation (CoV) of coating mass between tablets produced in a single batch, for which approximate closed form equations can be found in the literature [3]. Finally, the CoV is incorporated into the mechanistic tablet dissolution model to predict the drug release profile. The overarching goal of the proposed work is to maintain the quality of tablet coating and the drug release profile of the batch simultaneously during scale-up.

References:

1. Kaunisto, Erik, et al. 'Mechanistic modelling of drug release from polymer-coated and swelling and dissolving polymer matrix systems.' *International journal of pharmaceutics*1 (2011): 54-77.
2. am Ende, Mary Tanya, and Alfred Berchielli. 'A thermodynamic model for organic and aqueous tablet film coating.' *Pharmaceutical development and technology*1 (2005): 47-58.
3. Kalbag, Arjun, and Carl Wassgren. 'Inter-tablet coating variability: tablet residence time variability.' *Chemical Engineering Science*11 (2009): 2705-2717.

Author Disclosure: All authors are Sanofi employees and may hold shares and/or stock options in the company.

342f Decoupling Calibration and Coarse-Graining for DEM Simulation of Fine Particles

Lokeshwar Mahto, Jayanta Chakraborty, Jitendra Kumar, Anurag Tripathi, Maitraye Sen, William Ketterhagen

Calibration is an integral part of DEM simulations. However, the calibration methodology is still evolving. In a typical bulk calibration procedure, a set of bulk powder observables is determined by running a number of simulations for a set of DEM input parameters. Usually, coarse-grained particles are used in the simulations; hence, the calibrated parameters do not correspond to those for real particles. This feature reduces the generality of parameters. In this work we attempt to decouple coarse graining from calibration in order to obtain a more generally applicable set of parameters.

This methodology involves employing a surrogate system for the calibration of parameters. The surrogate system and the system of interest should have identical parameter sensitivity, which can be determined by running a set of quick simulations. The surrogate system should be such that the multitude of simulations needed for calibration may be conducted readily (e.g., using periodic BC).

Once the parameters for the resolved system (i.e. without coarse-graining) are identified using the surrogate system, the effect of coarse-graining on the parameters is studied using the same/another suitable surrogate system. Thus, the parameters corresponding to a coarse-grained system for a given coarse-grain ratio are obtained, and this final parameter is used for the system of interest to generate results that reproduce experimental results quantitatively. This methodology is shown in Fig. 1.

We demonstrate this method using Lactose Fast Flo 316 powder and a rotary tablet press feeder as the system of interest. A wedge-shaped hopper with a collection pan at the bottom was considered for the surrogate system to calibrate the properties of the resolved system, where the discharge rate and angle of repose were the observables. An Artificial Neural Network (ANN) based calibration methodology was used [1] to obtain the calibrated parameter space. The effect of coarse-graining on the parameters was studied using a simple shoe-die pair. Finally, the calibrated parameters corresponding to a given CG ratio are used for the DEM simulation of the tablet press feeder.

References:

[1] Forgber, T., Khinast, J. G., & Fink, E. (2022). A hybrid workflow for investigating wide DEM parameter spaces. Powder Technology, 404, 117440.

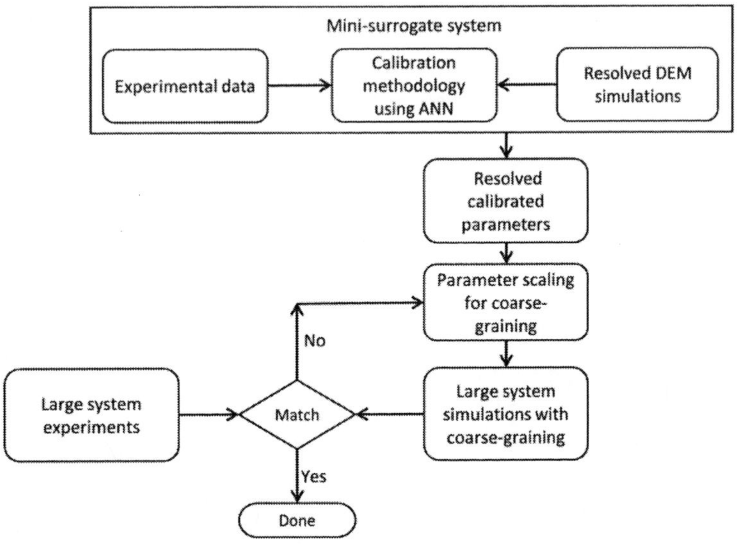

Figure 1. Overall calibration workflow

342g A Coupled Modeling Framework As a Soft Sensor for Monitoring the Drying of Active Pharmaceutical Ingredient Powders in Drug Substance Manufacturing Process Trains

Shashank Venkat Muddu, Ajinkya Pandit, Yi Wei, Yihui Yang, Wenlong Tang, Neda Nazemifard, Charles D. Papageorgiou, Richard Braatz, George Barbastathis, Allan Myerson

Active pharmaceutical ingredients (APIs) are the key components in pharmaceutical formulations that give the desired therapeutic effects. Several products are manufactured as oral solid dosage products in the form of either tablet or capsules. Prior to blending, granulation and milling of the final blend of APIs and other filler products, the APIs themselves are taken through several size engineering steps post synthesis in the form of crystallization, filtration, washing and drying. APIs being particulate products have several metrics associated with them such as a mean size, span, modalities and spread of the particle size distribution (PSD). The steps from filtration to the drying of the APIs can be carried out in agitated filter dryers (AFDs) which are versatile and widely used equipment in the industry.

Population balance models (PBMs) are the go-to tools for building simple yet reasonably mechanistic models for tracking the changes in a PSD of a particulate system during an engineering unit operation. Bulk heat transfer (HT) equation models are used to predict the heating and drying rates of the particulate API material. In this work, a PBM was coupled with a bulk HT model to calculate the drying rate and a real-time 'wetness' measurement of the particles. The coupled model was employed to give the PSD and wetness tracking of the bed during drying operations in an AFD for API development. The PBM consisted of various agglomeration and attrition phenomena occuring during the process which have been encoded as simple kinetic rate expressions as a function of important material properties of the components and the process variables relevant to an AFD operation. The coupled PBM-HT model framework has been proposed as a soft-sensor tool in manufacturing settings physical process analytical technology (PAT) tools cannot be employed due to safety regulations. The model would detect the endpoint for the drying and advise the engineer to end the operation a priori thereby leading to model-guided experimental studies.

The in silico model results generated have been compared to the various data generated using the in-house novel laser based PAT tool in our lab. The drying has been carried out for generic model API particles and anti-solvent combinations. The effect on the changes in PSDs have been reported along with the drying profiles. It is hoped that these model guided experimental studies would serve as building blocks for further development of AFD operations of APIs in pharmaceutical industry.

[1]: Pandit, Ajinkya, Qihang Zhang, Moo Sun Hong, Wenlong Tang, Charles D. Papageorgiou, Neda Nazemifard, Yihui Yang et al. 'Laser Speckle Probe for Monitoring Pharmaceutical Drying.' In 2023 AIChE Annual Meeting. AIChE, 2023.

[2]: Zhang, Q., Gamekkanda, J.C., Pandit, A. *et al.* Extracting particle size distribution from laser speckle with a physics-enhanced autocorrelation-based estimator (PEACE). *Nat Commun* 14, 1159 (2023). https://doi.org/10.1038/s41467-023-36816-2

342h Using Discrete Element Model Technique to Design a Densification Process for a Lyophilized Pharmaceutical Powder Formulation

Maitraye Sen, Callum Bruce

Powder flowability is an important attribute to consider while designing manufacturing processes for pharmaceutical solid dosage forms since it affects the die or capsule filling. Bulk powder properties, e.g., bulk density and particle size distribution affect the flow properties of the powder [1]. Pharmaceutical drug product formulations with low bulk densities or small particle size often have poor flow properties; hence the particles need to be engineered further so that the flowability improves.

This work presents a case study for densifying a lyophilized powder so that it can be efficiently encapsulated. Traditional methods for improving flow properties (e.g., wet granulation, roller compaction) are not applicable since lyophilized powders are sensitive to moisture and high temperature. Hence a mechanism using a densification piston was adapted. Examples of powder densification techniques using densification piston exists in other industries [2]. We adapted an approach where a densification piston was fitted at the outlet of a powder screw feeder. The piston applied a compressive force on the powder to form loose agglomerates of the powder particles and increase the powder bulk density, thus improving the powder flowability.

It is important to determine the magnitude of the compressive force to apply so that the bulk density of the powder is in an acceptable range that is amenable to encapsulation. The equipment configuration and settings determine the range of compressive forces that can be achieved. A discrete element model (DEM) was used to build the understanding and thus design the process. A qualitative DEM incorporating the Edinburgh Elastic Plastic Adhesion contact model [3] and representative model material properties was developed on EDEM (Altair EDEM). This work will highlight the model development and the outcome.

Acknowledgement: The work was done as part of a collaboration between Eli Lilly and Company and Altair EDEM when Callum Bruce was part of the EDEM team.

References

[1] D. Shah, K. Moravkar, D. Jha, V. Lonkar, P. Amin and S. Chalikwar, 'A concise summary of powder processing methodologies for flow enhancement,' Heliyon, vol. 9, no. 6, e16498, 2023.

[2] S. Obidzinski and J. Piekut, 'Influence of speed of densification piston and particle size of densified material on the value of densifying pressures and pellets density,' Journal of Research and Applications in Agricultural Engineering, vol. 60, no. 1, 63-68, 2015.

[3] S. Thakur, J. Morrissey, J. Sun, J. Chen and J. Ooi, 'Micromechanical analysis of cohesive granular materials using the discrete element method with an adhesive elasto-plastic contact model,' Granular Matter, vol. 16, 383-400, 2014.

149a Improving the Flowability of Lignocellulosic Biomass in Wedge-Shaped Hoppers Via External Vibrational Forcing: A DEM Study

Nicholas Deak, Yimin Lu, Hari Sitaraman, Jordan Klinger, Yidong Xia

The handling and transport of lignocellulosic biomass has emerged as a central topic in the production of biofuel because the low bulk density, polydispersity, and highly non-linear flowability (e.g. arching, ratholing in hoppers) of feedstock particles can impede downstream biorefinery processes and limit overall operations. Vibrational forcing has been utilized as a method for initializing and maintaining flow in hoppers, however, studies on this topic have mostly been limited to uniform granular materials. In this work, we use an in-house developed high-performance computing (HPC) compatible discrete element method solver to simulate the discharge of corn stover from a wedge hopper with and without the influence of external spatiotemporal forcing. The solver (BDEM, https://github.com/NREL/BDEM) includes models for representing complex particle shapes (glued and bonded-sphere models) and particle cohesion (liquid bridge and SJKR models), and has shown excellent strong-scaling performance on state-of-the-art CPU and GPU architectures for large problem sizes (> 100 M particles). Sinusoidal forcing is applied to the feedstock in both the vertical and horizontal directions, and the impacts of force amplitude and angular frequency are both explored. When amplitude and frequency were held constant, it was found that vertical and horizontal vibrations lead to similar average discharge rates, while the horizontal vibrations lead to lower variation in the discharge rate, reducing the intermittency that characterizes the vertical vibrations. In contrast with previous studies, the vibrational forcing amplitude is found to have a stronger impact on flow rate as compared with the angular frequency, as stronger vibrational forces are more effective in breaking the cohesive bonds that impede biomass feedstock flow. Finally, the hopper flow is visualized to show the flow patterns observed, providing deeper insight into the mechanisms by which vibrations help induce hopper flow. The horizontal vibrations induce flow near the walls of the hopper, resulting in an inverted funnel flow pattern, while the vertical vibrations result in funnel flow with a V-shaped valley characterizing the surface of the feedstock.

149c Drag Model Implementation for CFD-DEM Modeling of Dense Flow with Non-Spherical Particles in Ansys Rocky

Lucas B. Baggio, Lucilla Coelho de Almeida, Joao A. Aguirre Oliveira Jr., Emilio E. Paladino

Granular-fluid systems are often characterized by the complex interaction between solid particles and fluids that involve mass, momentum, and energy exchange. Such systems are found in various industrial processes, including sedimentation, filtration, and fluidized bed flows. The use of numerical simulations such as the coupled computational fluid dynamics (CFD) and discrete element method (DEM) model has become increasingly commonplace in modeling and predicting the behavior of granular-fluid systems as it allows quantification of process characteristics that are difficult to measure experimentally. With this approach, the fluid flow is typically treated as a continuous phase using averaged Navier-Stokes equations while the particles are modeled as a discrete phase. The coupling of the two phases relies on an accurate treatment of the drag force. For dilute flows, wherein particles can be assumed to be widely spaced apart and thus there exist no significant particle-particle interactions, empirical or semi-empirical drag laws accounting for the particle aspect ratio, orientation, or roughness are often used. Most granular-fluid systems, however, cannot be considered as dilute. In such dense flows, the effects of particle shape and particle-particle interactions become significant. These interactions may include collisions and the hydrodynamic screening effect that occurs when the drag experienced by each particle in a group of closely packed particles varies as a result of the fluid momentum being 'blocked' by the immediately adjacent particles.

This work concerns the selection, implementation, and validation of drag coefficient correlations for dense flows comprising non-spherical particles in Ansys Rocky and Ansys Fluent CFD-DEM coupling. The hybrid Ganser - Di Felice and Hölzer & Sommerfeld - Di Felice drag models were used. These models are formulated using the dense flow drag coefficient correlation presented by Di Felice for spherical particles as the basis, but using the drag coefficient correlations proposed by Ganser and Hölzer & Sommerfeld for application to non-spherical particles.

The drag models were integrated into the numerical model using the Rocky Solver Application Programming Interface (API). The experimental data in Volmari et al. was used to validate the model. As shown in Fig. 1(a), a fluidized bed with sphero-cylindrical particles in a square cross-sectional area with superficial velocities ranging from 0.4 m/s to 2.2 m/s was modeled to match the experimental setup described in the aforementioned reference. The numerical results obtained when using the hybrid drag laws provided a more accurate pressure drop prediction when compared to the result obtained using the original Di Felice drag law model in Ansys Rocky for a wide range of superficial velocities, as shown in Fig. 1(b). In addition, an overprediction of the minimum fluidization velocity results when the particle shape is not taken into account when computing the drag force, as seen in Fig. 1(c).

These results demonstrate that combining the Di Felice drag model and the Ganser and Hölzer & Sommerfeld drag coefficient correlations provides a more accurate estimation of drag forces in dense non-spherical particle flows compared to the standard Di Felice model. Thus, its implementation in the Ansys software framework when modeling granular-fluid systems must be considered.

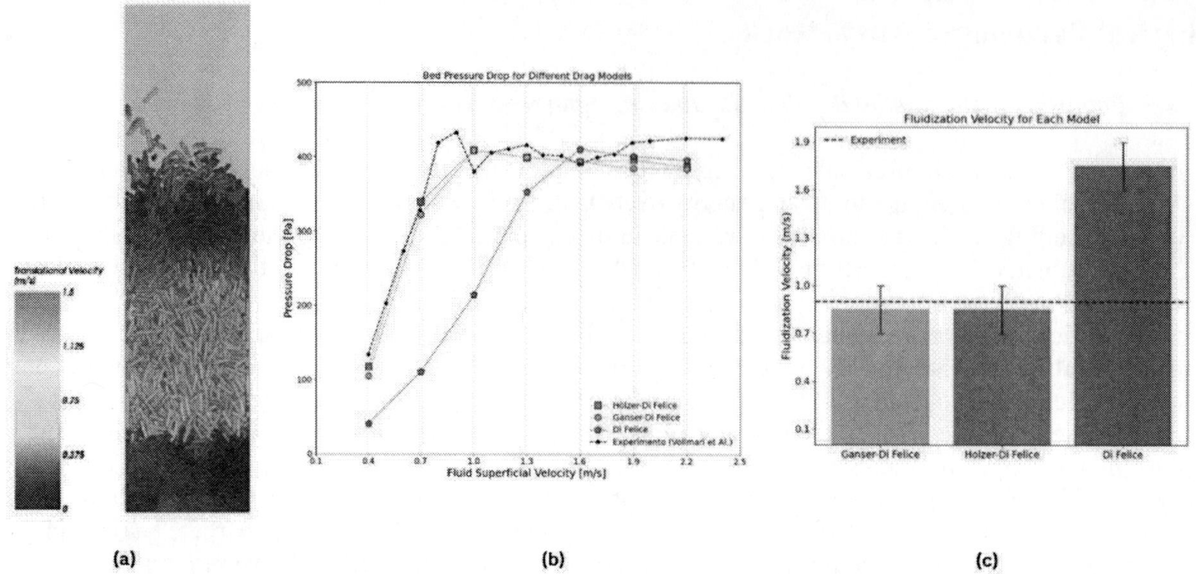

Figure 1. (a) Fluidized bed with sphero-cylinder particles colored by the particles' translational velocities, (b) pressure drops in the bed, and (c) minimum fluidization velocity prediction by each drag model used in the simulation.

149d Population Balance Model and Deep-Learning Model for Predicting Biomass Particle Size Reduction in Pilot-Scale Knife Mill

Yidong Xia, Minglei Lu, Tiasha Bhattacharjee, Jordan Klinger, Zhen Li

An extended population balance model (PBM) and a deep learning-based enhanced deep neural operator (DNO+) model are introduced for predicting particle size distribution (PSD) of comminuted biomass through a large knife mill. Experimental tests using corn stalks with varied moisture contents, mill blade speeds, and discharge screen sizes are conducted to support model development. A novel mechanism in the extended PBM allows for including additional input parameters such as moisture content, which is not possible in the original PBM. The DNO+ model can include influencing factors of different data types such as moisture content and discharge screen size, which significantly extends the engineering applicability of the standard DNO model that only admits feed PSD and outcome PSD. Test results show that both models are remarkably accurate in the calibration or training parameter space and can be used as surrogate models to provide effective guidance for biomass preprocessing design.

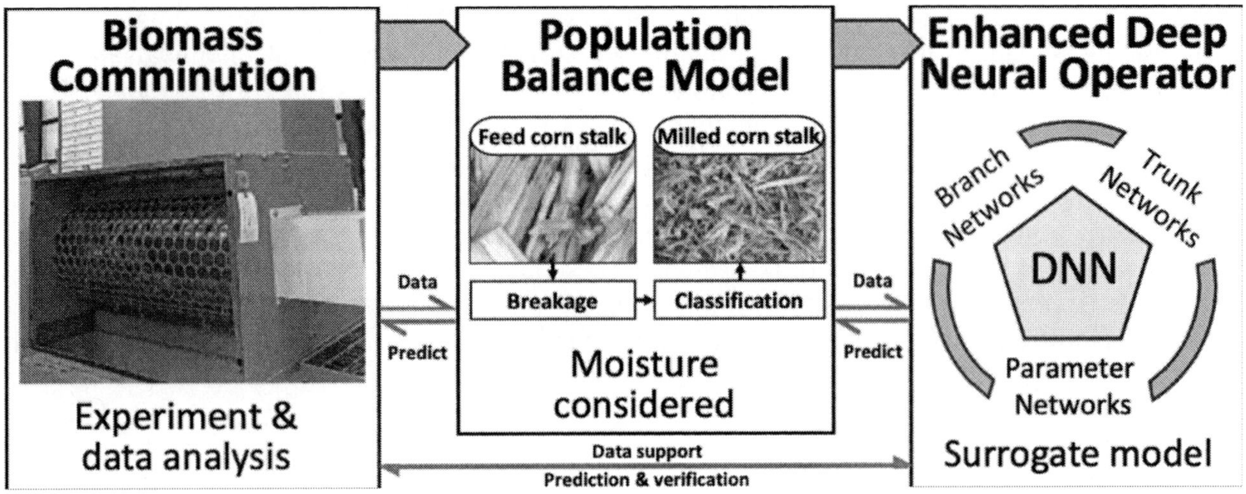

87a Mechanistic Investigation of Preprocessing Specifications Impact on Flow Related Issues in Biomass Handling Systems Using Experiment and Discrete Particle Simulations

Ahmed Hamed, Yidong Xia, Nepu Saha, Jordan Klinger

Biomass is an abundant, versatile source of renewable energy that can be used to generate power on demand and produce bio-based fuels, chemicals, and materials. Harnessing the full potential of bioenergy as a strategic player in the transition to carbon-neutral energy sources is contingent upon the deployment of this technology at commercial scale becoming economically viable. To achieve this goal, efficient processes need to be developed for different unit operations along the bioenergy value chain, from harvesting the raw biomass materials all the way up to their lifetime in the biorefinery. The unique technical challenges associated with the life cycle of biomass feedstocks is mainly attributed to the undesirable characteristics of their raw form, e.g., high variability, high moisture and oxygen content, low bulk density, low calorific value, and hydrophilic nature. Accordingly, biomass preprocessing is required to produce high-quality feedstocks using mechanical, chemical, and thermal processes that change the physical properties and chemical composition of the raw materials. The complex nature of biomass feedstocks, including their irregular shapes, variable moisture content, and heterogeneous composition, contributes to their poor flowability characteristics. As a result, biomass particles often exhibit non-uniform flow patterns, arching phenomena, and cohesive behavior, leading to flow obstructions, materials jamming, clogging, erratic flow, particle segregation, and loss of live storage volumes in storage and handling systems. These flow-related issues not only disrupt material flow but also increase energy consumption, maintenance costs, and operational downtime in biomass processing facilities. Understanding the underlying mechanisms of biomass poor granular flowability is crucial for optimizing the design and operation of biomass handling facilities and mitigating flow-related issues.

The flow behavior of particulate biomass materials is typically intermediate between solids and liquids. Individual particles and bulk biomass have complex viscoelastic deformation under stress. Under short-term low loading, the yield stress can withstand imposed stresses without any significant deformation up to a threshold value. Stresses above this threshold value causes permanent deformation in a complex fashion that involves localized elastic and plastic strains, arising from the discrete nature of the particles, and exhibits nonlinear flow response to the applied external forces. The threshold stress required for the onset of permanent deformation depends strongly upon the inter-particle friction and is very sensitive to the stress and deformation histories. Difficulties associated with the handling and feeding of biomass feedstocks are not originating only from the traditional rheological properties such as bulk density, moisture content, compressibility, cohesive and adhesive strengths, as it is also affected by the chemical composition of individual particles, internal microstructure, entrapped gases, and temperature. For example, while some anatomical fractions of loblolly pine (bark and needle) exhibited direct relation between the moisture content and the rheological properties, no influence was observed for other fractions (stem and whole). In addition, the impact of the moisture content on the rheological properties was only present for the smaller particle size of bark and needle. Moreover, same-size particles from different fractions showed different flowability. This demonstrates the inherent variability of the biomass feedstocks attributes and their impact on the final product quality. So, it is extremely important to thoroughly quantify the impact of different material attributes on the performance of different unit operations along the supply chain, to understand the interplay between them and assess the critical parameters.

Under dynamic loading conditions, the granular flow behavior can be classified into three different regimes, namely, 1) quasi-static shearing, 2) inertial, and 3) intermediate transitional flows. While the quasi-static shearing flow regime represents one extreme that occurs in dense and confined granular assemblies, the inertial flow is the opposite extreme that is typically encountered in the dilute limit where the granular particles are loose and interacting mainly through uncorrelated binary collisions. Theoretical formulation of elastoplastic solids and the kinetic theory of dense gases can be adequately applied to the two different regimes, respectively. On the other hand, the intermediate transitional flow regime is so far still poorly understood. Thus, numerical experiments are needed to complement theoretical prediction and aid in bridging the gap between the microscopic properties and the macroscopic behavior of granular materials. The present research explores the challenges associated with the poor granular flowability of biomass feedstocks in different materials handling and feeding unit operations along with validation efforts using experiment. This includes screw feeder, roll screen, hopper, and compressive screw pump systems. Applications of several Discrete Element Method (DEM) models are sought to quantify the flowability of granular biomass feedstocks (using mass throughput and specific energy consumption as benchmarking metrics) and study the impact of different comminution techniques and preprocessing specifications on the performance of handling systems under consideration. Specifically, following quality-by-design approach, different critical material attributes (particle size, particle size distribution, particle shape, and moisture content) and critical process parameters were investigated. Effects of factors such as particle properties, inter-particle interactions and interlocking, wall friction, and surface roughness on jamming phenomena and transitions between flow regimes are simulated. Simulation results shed light on the fundamental mechanisms leading to jamming events and help advance the modeling of flow-related issues in industrial applications. This will enable the development of mitigating strategies and innovative solutions for optimizing biomass handling systems, improving process efficiency, and promoting sustainable bioenergy production.

87b Variability of Sphere- and Tetrapod-Based DEM-Simulations for Poorly Flowing Materials

Lukas Maier, Michael Mitterlindner, Gregor Fasching, Stefan Radl

In the realm of chemical engineering applications, it is imperative to embrace innovative solutions like residue-derived fuels, battery recycling, and biomass utilization. It is clear, that approaches like these pave the way towards a more sustainable future. Thus, the modelling of such, rather poorly flowing, systems is of crucial importance. However, the respective primary particles are complex shaped and therefore not well suited as a basis for a useful DEM-simulation approach. This brings forth the concept of treating them as parcels [1]. In the current approach, tetrahedral multi-spheres with non-touching spheres ("tetrapods") are used as a basic geometric concept for those parcels.

Our current work investigates the variability of predictions using a tetrapod approach [2], as well as using conventional spheres. We focus on a simple draw-down test, as shown in Figure 1, and study the effect of the filling procedure's parameters on the models prediction [3]. Furthermore, the effect of using distributions for (i) the tetrapod properties and (ii) the force-parameters in the contact model are investigated in the spirit of O'Sullivan et al. [4]. Specifically, we are successful to describe fluctuations occurring naturally in our lab experiments (and in real world applications), which are due to the nonideality of such systems. Finally, the effect of parcel size (which is known to be non-trivial as shown by Tausendschön et al. [5]) is quantified, and an attempt is made to calibrate the DEM simulator also for correct prediction of the fluctuations seen in our experiments. Using our novel approach, systems with poorly flowing materials and strong nonidealities can be quantitatively described.

[1] S. Radl, C. Radeke, J. G. Khinast, and S. Sundaresan, "Parcel-Based Approach For The Simulation Of Gas-Particle Flows" *8th Interantional Conference on CFD in Oil & Gas, Metallurgical and Process Industries*, no. June, 2011.

[2] S. Radl, H. Benabchiasli, G. Fasching, M. Mitterlindner, and M. Salehi, "How CFD-DEM Simulations benefit from Machine Learning", Editorial to the Proceedings of 11th European Conference on Industrial Furnaces and Boilers, INFUB-14, 2-5 Apr. 2024.

[3] J. Leak and D. Barreto, "A DEM study on the effect of inherent variability of assemblies of spherical particles", doi: 10.53243/NUMGE2023-388.

[4] C. O'Sullivan, J. D. Bray, and M. F. Riemer, "Influence of Particle Shape and Surface Friction Variability on Response of Rod-Shaped Particulate Media", doi: 10.1061/ASCE0733-93992002128:111182.

[5] J. Tausendschön, J. Kolehmainen, S. Sundaresan, and S. Radl, "Coarse graining Euler-Lagrange simulations of cohesive particle fluidization" *Powder Technol*, vol. 364, pp. 167–182, Mar. 2020, doi: 10.1016/J.POWTEC.2020.01.056.

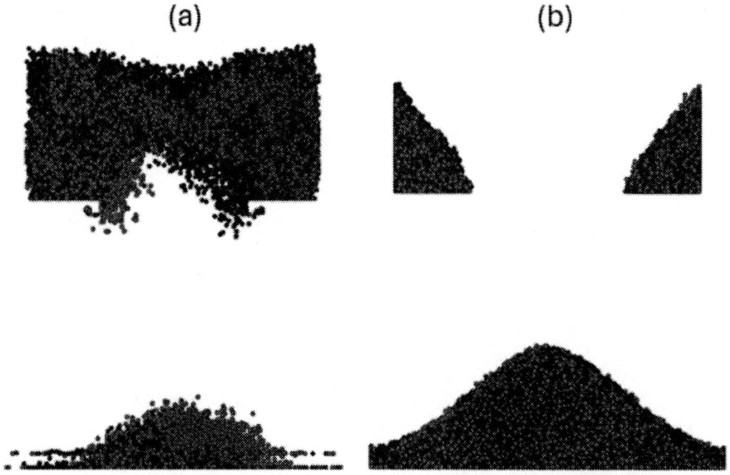

Figure 1: Different filling parameters indicated by red and black particles result in (a) a very different contour of the heap for tetrapod parcels and (b) the same contour of the heap for spherical particles

87c A Machine Learning Assisted Hopper Flow Design for Handling Granular Biomass Materials

Wencheng Jin, Abdallah Ikbarieh, Yumeng Zhao, Xuyang Li, Nepu Saha, Jordan Klinger, Yidong Xia

The potential of biomass-derived energy is significant, yet its realization is hindered by handling and feeding challenges associated with granular biomass feedstock. For instance, the prevalent issues in biorefinery hoppers include arching, rate hole formation, and avalanche flow. These challenges stem from gaps in understanding the relationships between material properties, hopper design factors and flow performance.

In this study, we integrate physical experiments, validated numerical simulations, and data augmentation techniques to develop a machine learning-assisted hopper design tailored for flowing granular woody biomass materials. Initially, we employed experimental data from cyclic axial compression and ring shear tests to train two deep-learning networks: a sequential model utilizing Gated Recurrent Units as the encoder and feed-forward neural networks as the decoder, and an incremental model using feed-forward neural networks. Upon validation, these networks were utilized to enhance the laboratory-measured stress-strain behavior of pine chips across various moisture contents and mean particle sizes.

Subsequently, we employed a meticulous calibration process to adapt a modified hypoplastic model capable of accurately representing the intricate flow behavior of granular biomass materials. These calibrated constitutive models, in conjunction with a variety of initial packing configurations and hopper operational conditions (e.g., outlet opening size, wall friction, inclination), facilitated hopper flow simulations using a validated Smoothed Particle Hydrodynamics code. For each simulation, we defined and computed a set of performance indices, such as average flow rate, clogging potential, flow smoothness, and flow pattern coefficients, to quantitatively assess flow performance.

Finally, a feed-forward neural network was trained and validated to establish correlations between material attributes, initial packing, hopper operation parameters, and flow performance indices. This neural network serves as a comprehensive design guideline for effectively handling milled woody biomass materials in hoppers, benefiting stakeholders in biorefinery and equipment manufacturing sectors.

87d Kinetic Theory Closures with Finite Contact Time for Electrostatic Charge Generation and Transport

Saykat Poddar, Manjil Ray, Alberto Passalacqua

Electrostatic charging is a critical phenomenon in systems that manufacture insulating particles because particle collisions with walls or among themselves lead to significant charge accumulation and electric potential difference between the particles and the ground, affecting both the functionality of the production systems and the safety of their operations.

Charge transport models have been developed over the past few years [1–5] to describe the evolution of electrostatic charge in such systems. Our previous work [1,2] focused on Eulerian-Eulerian multi-fluid models with kinetic theory closures for the particulate phase due to their wide adoption to simulate devices of practical interest. Such models have demonstrated charge separation and particle buildup at reactor walls. However, they do not account for finite collision times, between particles. Including such effects modifies the collision frequency, affecting the dissipation of granular energy and, consequently, the fluid dynamic properties of the disperse phase and the rate of charge separation due to particle-particle and particle-wall collisions. Considering finite contact time in the Eulerian model closures for electrostatic charging also addresses an inconstancy present in previously formulated models, where Hertzian collisions with finite contact time are considered in the charge separation model [6], while kinetic theory closures for the hydrodynamic properties of the disperse phase are determined with standard kinetic theory closures [7], assuming instantaneous collisions (t_c=0).

We consider the work of Berzi and Jenkins [8] and leverage their findings to formulate closures for electrostatic charging accounting for finite contact time, whose effect on the prediction of electrostatic charging is then evaluated by comparing the predictions obtained with the new closures to those provided by previously derived closures considering instantaneous collisions.

References

[1] M. Ray, F. Chowdhury, A. Sowinski, P. Mehrani, A. Passalacqua, An Euler-Euler model for mono-dispersed gas-particle flows incorporating electrostatic charging due to particle-wall and particle-particle collisions, Chem. Eng. Sci. 197 (2019) 327–344. https://doi.org/10.1016/j.ces.2018.12.028.

[2] M. Ray, F. Chowdhury, A. Sowinski, P. Mehrani, A. Passalacqua, Eulerian modeling of charge transport in bi-disperse particulate flows due to triboelectrification, Phys. Fluids 32 (2020) 023302. https://doi.org/10.1063/1.5140473.

[3] J. Kolehmainen, A. Ozel, S. Sundaresan, Eulerian modelling of gas–solid flows with triboelectric charging, J. Fluid Mech. 848 (2018) 340–369. https://doi.org/10.1017/jfm.2018.361.

[4] J. Kolehmainen, A. Ozel, C.M. Boyce, S. Sundaresan, A hybrid approach to computing electrostatic forces in fluidized beds of charged particles, AIChE J. 62 (2016) 2282–2295. https://doi.org/10.1002/aic.15279.

[5] L. Ceresiat, J. Kolehmainen, A. Ozel, Charge Transport Equation for Bidisperse Rapid Granular Flows with Nonequipartitioned Fluctuating Kinetic Energy, ArXiv200904503 Phys. (2020).

http://arxiv.org/abs/2009.04503 (accessed April 24, 2021).

[6] J.C. Laurentie, P. Traoré, L. Dascalescu, Discrete element modeling of triboelectric charging of insulating materials in vibrated granular beds, J. Electrost. 71 (2013) 951–957. https://doi.org/10.1016/j.elstat.2013.08.001.

[7] J.T. Jenkins, S.B. Savage, A theory of the rapid flow of identical, smooth, nearly elastic spherical particles, J. Fluid Mech. (1983) 187–202.

[8] D. Berzi, J.T. Jenkins, Steady shearing flows of deformable, inelastic spheres, Soft Matter 11 (2015) 4799–4808. https://doi.org/10.1039/C5SM00337G.

421a A Novel Predictive Approach to Mitigate Tablet Film Coating Scuffing

William Ketterhagen, Andreas Thomann, Florian Ries, Andreas Gryczke

Tablet film coating processes are often conducted to improve the aesthetics of the dosage form and improve robustness for downstream processing. In certain cases, however, the film coating process itself can be implicated in visual defects in the form of small gray or black blemishes on the surface of the tablets – so called tablet scuffing. While the causes of scuffing have been investigated by several groups, mitigation of tablet scuffing in practice can still be a burdensome and empirical process. Further, scuffing usually only becomes apparent in large scale processes, further increasing the resource consumption to resolve this issue.

In this work, a novel predictive approach is developed and applied to identify ways to alleviate tablet scuffing observed in commercial scale pan coating processes. The approach combines computational simulation of the coating process using the discrete element method (DEM) and a novel, lab-scale characterization of scuffing susceptibility. The DEM simulations capture the scuffing energy imparted to the tablets for various process conditions and coating pan geometries. Further, the DEM simulations also provide valuable insight into the genesis of scuffing, i.e., where and when scuffing occurs in the coating pan, which is in alignment with historical observations. Characterization of scuffing susceptibility in the lab allows for the development of a direct relation between the scuffing energy and the manifestation of observable scuffing. This characterization accounts for specifics of the materials involved, e.g., coating formulation and steel finish, that may affect the extent of scuffing.

This new predictive approach now enables quantitative prediction of tablet scuffing using only a sample of a few coated tablets. Various process-related mitigation approaches can be studied in silico, thereby reducing large scale development trials previously required. Further, the approach can also be applied to future development programs to proactively design coating processes with minimal tablet scuffing risk.

Disclosures

AbbVie contributed to the design; participated in collection, analysis, and interpretation of data; and in writing, reviewing, and approval of the final version. AbbVie sponsored and funded the study. All authors are current employees of AbbVie and may own AbbVie stock.

421b The Effect of Wall Friction and Baffles on Particle Flow and Heat Transfer in a Rotary Drum

Carlin Leung, Marcella Alves, Elaheh Ardalani, William Borghard, Nina C. Shapley, Alberto Cuitino, Benjamin Glasser

Rotary drums are a common equipment used in several industries such as in catalysts, pharmaceutical, chemical, and the food industry. They are used for thermal treatment for a wide range of materials. Better understanding of particle flow and heat transfer can greatly improve the product quality and reduce wasted energy and materials. It is necessary to uniformly increase the material temperature for good product quality. Typically, rotary drums are made of steel, but can be made of cast iron or ceramic, each having significantly different textures and surface roughness. Surface roughness can have a significant effect on heat transfer and temperature uniformity on the material. Studies have been done on mixing of materials in rotary drums. However, to the best of the authors' knowledge there are no studies focusing on heat transfer based on equipment surface roughness in a rotary drum.

Understanding the relationship of equipment surface roughness on heating time is important for predicting processing time in real-world applications. In this work, simulations using the discrete element method (DEM) and the Hertz-Mindlin contact model were conducted to better understand the effect of surface roughness (coefficient of static friction) on heat transfer for non-cohesive particles. The model accounts for particle-to-wall and particle-to-particle interactions. The coefficient of static friction between particle and wall and the number of baffles in the rotary drum were investigated. We found that static friction has a significant effect on thermal time and temperature uniformity. Depending on the static friction tested, two flow regimes were exhibited: Slipping and rolling regime. At low static frictions, thermal time is high and temperature uniformity is poor, while increasing static friction, decreases thermal time and improves temperature uniformity. Particle tracking and total kinetic energy of the particles were investigated to further understand the effect of static friction on heat transfer and flow patterns. Baffles were added in the simulations to investigate their effect on heat transfer for different values of static friction. We found that adding baffles for low static friction cases greatly improves heat transfer and temperature uniformity, but consequently increases the total kinetic energy of the particles, which can lead to attrition. This work provides quantitative findings on how equipment design affects heat transfer and temperature uniformity in a rotary drum.

421c Application of Multi-Level Coarse-Grained Model for DEM Heat Transfer Simulation

Tarun De, Ashok Das, Daniele Marchisio, Gianluca Boccardo

In the evolving landscape of heat transfer simulation, the discrete element method (DEM) has emerged as a pivotal tool for predicting the thermal behaviour of granular materials. Building upon a recent advancement in multi-level coarse-graining (MCG) techniques for DEM simulations, this study introduces novel scaling laws for thermal conductivity and thermal capacity tailored for coarse-grained particles. These scaling laws are integral to the proposed multi-level coarse-grained simulation framework, enabling the efficient simulation of heat transfer processes in systems comprising a large number of particles. By leveraging the particle location-based method for multi-level coarse-graining, we delineate zones of varying resolution within the simulation domain, allowing for dynamic adjustment of particle granularity and preserving overall thermal temperature. This approach not only preserves the computational efficiency afforded by coarse-graining but also enhances the accuracy of thermal simulations by accounting for the heterogeneity in thermal properties across different scales. In our study, we validated our model through a hopper discharge simulation, featuring cold and hot temperature particles on the left and right halves, respectively. We compared the thermal temperature and heat flux distributions across the bed height and among discharged particles between MCG simulations with scaled thermal parameters and fully resolved simulations, finding strong agreement. Additionally, we examined MCG simulations without scaled thermal parameters against resolved simulations, highlighting the necessity of accurate thermal parameter scaling for reliable simulation outcomes. This comparison highlights the significance of proper parameter scaling in achieving accurate simulation outcomes. This study paves the way for more accurate and computationally feasible simulations of industrial-scale heat transfer applications.

*This research is partially financially supported by ICSC—Centro Nazionale di Ricerca in High Performance Computing, Big Data and Quantum Computing, funded by European Union - NextGenerationEU.

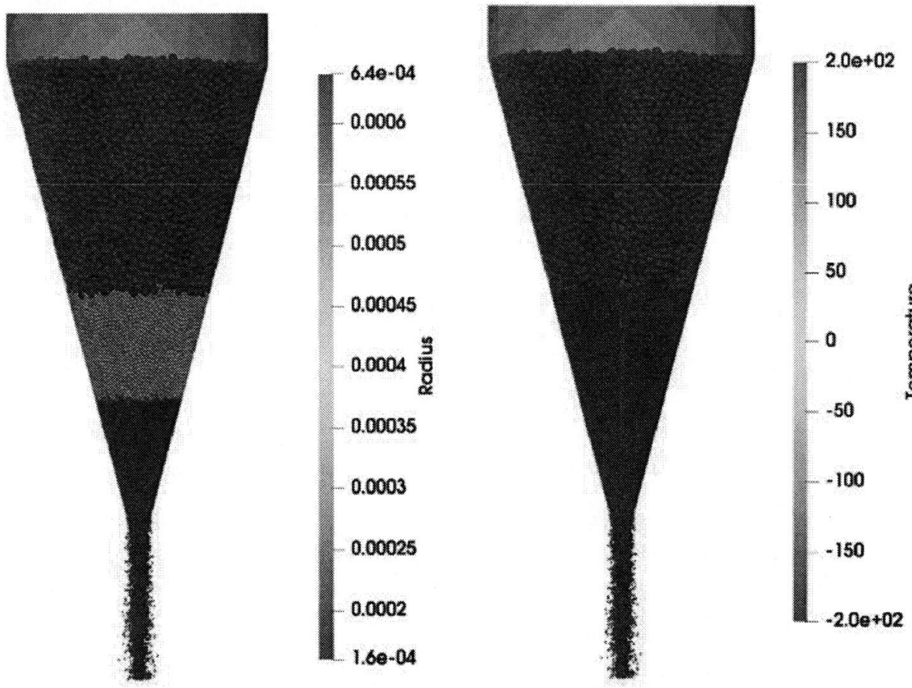

(a) Different Radii during MCG Simulation.

(b) Temperature during MCG Simulation.

Figure 1: Snapshots during MCG simulation.

421d Thermal Modeling of Poorly Flowing Bulk Solids Using Tetrapods Using the DEM

Michael Mitterlindner, Lukas Maier, Gregor Fasching, Stefan Radl

The simulation of poorly flowing (cohesive and strongly interlocking) bulk solids will play a crucial role in the near future, providing a pathway towards efficient utilization of valuable resources and mitigation of environmental impacts. Biomass utilization and different recycling processes, e.g., for batteries or polymers, are examples. Our present contribution focuses on simulating the handling of shredded nickel-metal hydride (Ni-MH) batteries during a typical recycling process: after a shredding step, the material must be conveyed and cooled prior to storage.

The modeling process is challenging due to (i) the cohesive nature of the material (mainly caused by interlocking flakes, fine particles, and liquid bridges), (ii) large differences in the material behavior when stress is applied (ranging from low stress flow to high pressure compaction), and (iii) the low effective heat conductivity of the material. To address these challenges, we propose a novel parcel-based DEM approach [1, 2] using tetrahedral multi-spheres (i.e., "tetrapods", see Figure 1). Specifically, two key improvements were implemented in the tool LIGGGHTS [3]:

1) Finite intra-tetrapod heat conduction: this feature overcomes the unphysical infinite heat rate transfer problem, present in the classical multi-sphere approach. Our 'intra multi-sphere thermal conductivity' model predicts the exchanged heat between individual spheres that constitute the tetrapod. By adjusting this conductivity value, heat transfer within a tetrapod can be tuned to better reflect the presence of air gaps and match the effective bulk conductivity of shredded battery material.

2) Flexible tetrapods: this feature allows individual spheres within a tetrapod to displace towards the center of mass. While rigid tetrapods (i.e., the vanilla multi-sphere implementation) with a simple cohesion model already perform well in low stress scenarios (e.g., a draw-down test, see Figure 2), a compaction process requires this more complex modeling technique. For example, our plastic deformation model of the tetrapods can be used to represent the inherent tendency of the material to remain compacted after the stress has been released.

[1] S. Radl, C. Radeke, J. G. Khinast, and S. Sundaresan, "Parcel-Based Approach For The Simulation Of Gas-Particle Flows," *8th Interantional Conf. CFD Oil Gas, Metall. Process Ind.*, no. June, pp. 1–10, 2011.

[2] J. Tausendschön, J. Kolehmainen, S. Sundaresan, and S. Radl, "Coarse graining Euler-Lagrange simulations of cohesive particle fluidization," *Powder Technol.*, vol. 364, pp. 167–182, 2020, doi: 10.1016/j.powtec.2020.01.056.

[3] C. Kloss, C. Goniva, A. Hager, S. Amberger, and S. Pirker, "Models, algorithms and validation for opensource DEM and CFD-DEM," *Prog. Comput. Fluid Dyn.*, vol. 12, no. 2–3, pp. 140–152, 2012, doi: 10.1504/PCFD.2012.047457.

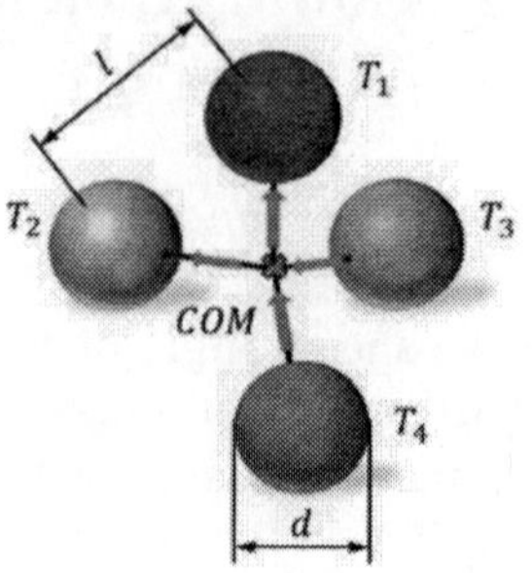

Figure 1: Conceptual sketch of a tetrapod used for thermal modeling.

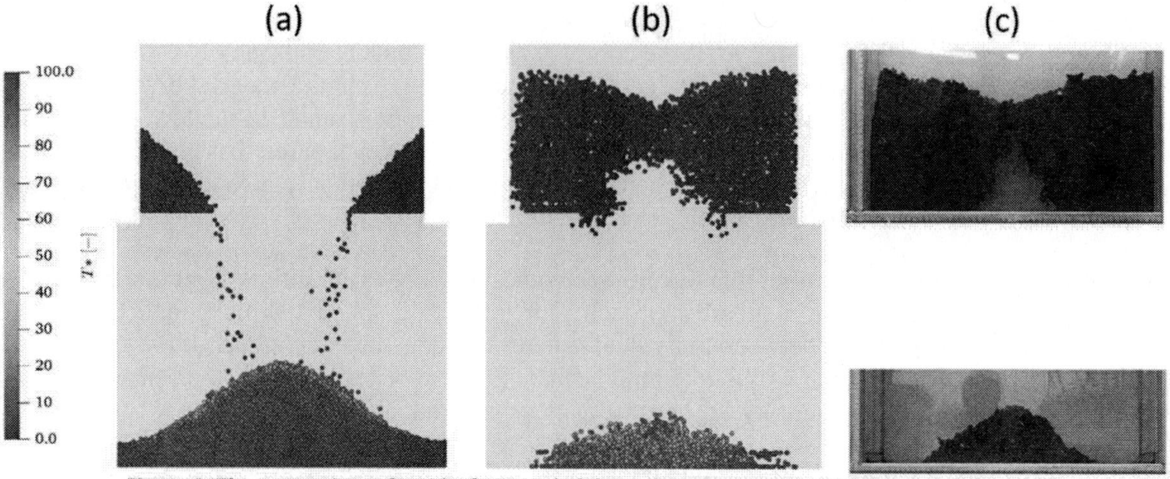

Figure 2: The comparison of results for a cooled draw-down test between simulations using only spheres (panel a; colors indicate particle temperatures), simulations using tetrapods (panel b), and experiments (panel c).

421e Caking Problems and the Role of Flow Aids – a Systematic Approach

Kerry Johanson

One problem of working with bulk materials is that sometimes they just do not handle well after storage. We call this problem caking. The bulk material is placed into a container, bin, hopper, feeder, or other process vessel, and – over time – the material becomes very cohesive and plugs the process equipment. Sometimes this time is long and sometimes it is very short. So, what is the solution? One potential solution is to use flow aids so as to limit the build-up of unconfined yield strength during storage. From a continuum point of view, unconfined yield strength is defined as the major principal stress that causes a bulk material to yield in shear while in an unconfined state after it has experienced a consolidation stress causing it to compact. From a particle viewpoint, during a shear event adjacent particles are in the process of two distinct actions. Some adjacent particles slide past each other, causing frictional resistance forces. Some adjacent particles adhere to each other, and during a shear event, these particles pull apart and rupture the adhesion connection. So, the bulk unconfined yield strength (a fundamental continuum property) is a function of the collection of all the friction forces as well as all the adhesion forces between adjacent particles. All adhesion force laws depend on how close the adhesion surfaces are.

With this in mind, the traditional simplistic view of flow aids suggests that the addition of a flow aid works because the flow aid particles fit between the larger particles and separate these larger particles, thereby reducing the adhesion forces between particles and decreasing the bulk unconfined yield strength. So, the use of flow aids to mitigate cohesive flow problems should be solved by simply computing the coverage of a particle by finer flow aid particles and then using adhesion laws to determine the reduction in adhesion forced because of the extra distance between particles caused by the coated particle surfaces. Based on this simple viewpoint one would expect that it would not matter what type of flow aid particles are used to coat the bulk particle surfaces. However, real life observation is that the type of particles on the surface of a bulk material significantly affects the reduction in bulk cohesive flow properties. In addition, the cause of the bulk cohesive strength in caking environments plays a role as well. In cases where caking is induced by recrystallization effects, the degree of hydrophobicity of the flow aid plays a significant role in the reduction (or not) of strength. The particle size and size distribution of the flow aid relative to the bulk solid is also a factor. Likewise, in cases where caking is caused by sintering effects, the size of the flow aid and the softness of the bulk particle surface play an important role in the reduction of caking problems with the use of flow aids.

Caking is inherently a mechanistic problem, meaning that it has one or more causes that may happen at the same time. Each caking mechanism has an associated time constant and behavior. Each caking mechanism will also be sensitive to something about the environment that initiates the caking and determines the rate at which the problem occurs. Caking depends on the chemistry of the material, the size and shape of the particles, the particle size distribution, the geometry of the process vessel, and the environmental changes that the process vessel may experience over time. Often caking events can be avoided by using a flow aid to prevent or limit the caking causes. However, sometimes the use of a flow aid does not work. In this event engineers, process managers, or formulation developers need to know the root cause of the caking issue before an alternate flow aid can be effectively chosen. In some cases, the flow aids chosen may make the material more prone to problems. The question is – WHY? The ultimate goal is to understand why flow aids sometimes do not work but, instead, join the dark side and cause even more problems.

421f Flow Pattern Evolution Law and Mechanism Analysis of Biomass Particles in Horizontal Pipe

Hui Du, Haifeng Lu, Xiaolei Guo, Haifeng Liu

The analysis of flow patterns in pneumatic conveying processes is crucial for optimizing the design and operation of conveying systems. The pneumatic conveying flow pattern primarily encompasses three aspects: solid particle movement, gas phase flow, and gas-solid interaction. Electrical Capacitance Tomography (ECT) was an effective and advanced measure method that was used to research flow patterns in rice husk powder dense-phase pneumatic conveying with a solid–gas ratio of 60–100 kg/kg. The evolution of the flow pattern in the pipe and the concentration distribution in the cross-section were characterized by ECT. The impact of gas velocity on solid deposition, gas phase flow, and the gas-solid interaction area in the pipeline was analyzed. Plug flow, fluidized flow, annular flow and stratified flow were observed in the horizontal pneumatic conveying. The relationship between Reynolds number and Archimedes number was used to predict the transition of these flow patterns. By examining the gas-solid interaction, the physical mechanism of the biomass flow pattern in the pipeline was explored, and a gas-solid interaction model was established to comprehend the operational principles of the pneumatic conveying system. This model offered a theoretical foundation for optimizing the design and operation of the system.

653a Development of Spray-Dried Carriers for *in-Situ* Production of Volatile Phytochemicals

Ondrej Kaspar, Lucie Maskova, Viola Tokarova

The emergence of MDR bacterial strains and the lack of new antibiotics discovered in the past 30 years, known as the discovery void, are major risks to modern society. Less than 100 years after the breakthrough discovery of penicillin, we are facing the grim scenario that even common superficial injuries or routine surgical procedures may become dangerous and life-threatening again. Many strategies have been developed, but most fail due to the continual adaptation of bacteria. Therefore, an alternative approach and out-of-the-box solutions are needed. For instance, new antibiotics were discovered with the assistance of AI machine learning [1]. Nevertheless, due to practical and financial difficulties associated with developing new antibiotics, most pharmaceutical companies already shifted their attention to cost-effective and more lucrative products. On the other hand, the development of bacterial resistance against potent but highly reactive and unstable thiosulfinates found in Allium plants is more than 1000 times slower compared to commonly used antibiotics [2]. A combination of the short half-life, high reactivity (oxidation of thiol groups), and non-specificity to particular proteins are reasons why most bacteria cannot effectively cope and develop defensive mechanisms.

Garlic (*Allium sativum*) is a typical example of a compartmentalized natural system, where the highly potent but unstable compound allicin is enzymatically formed from its precursor alliin only when the inner cellular structure is compromised. This concept makes allicin an everlasting bactericide since it is locally formed when needed, leaving no permanent traces of its active form in the environment after application. Alliin, a substance found in intact garlic cells, is not the only sulfoxide involved in the natural self-defence of the genus Allium. Other plants like onion or leek have chemical analogues to alliin, e.g., isoalliin, methiin, propiin, etc. These plant-specific compounds with similar chemical backbones share identical reaction mechanisms leading to the enzymatic production of various thiosulfinates. It has been reported that allicin-like analogues exhibit similar properties. However, their volatility and partition coefficient depend on a particular sulfoxide's molecular weight and character [3]. Moreover, allicin analogues' physicochemical properties, biological activity on various bacterial strains, and cytotoxicity towards different human cell lines are yet to be explored.

Allicin, the substance responsible for garlic's beneficial properties, is one of the most bioactive compounds found in nature. However, allicin's high instability, which is simultaneously responsible for its persistent biological activity, has always been considered a critical weakness preventing further practical and broader use. On the other hand, it has been reported that allicin in very low concentrations can increase the therapeutic effect of FDA-approved antibiotics, causing temporal permeabilization of bacterial membranes [4]. For all these reasons, many research groups around the world are focusing on allicin and its practical applications due to its numerous benefits.

Controlled *in-situ* synthesis of allicin near a target site may be the feasible way to (re)utilize its full potential. For that, it is essential to physically separate and stabilize allicin's precursors (enzyme and substrate) and introduce a controlling mechanism to harness the beneficial properties of allicin. Driven by the urgency for novel approaches to prevent pandemic scenarios caused by multi-drug resistant bacteria, we propose spray-dried carriers suitable for a topical or pulmonary application, *in-situ* formation, and controlled release of antibacterial thiosulfinates. This approach can significantly extend the limited palette of currently used antibacterial compounds for thiosulfinates, overcoming the lack of antibiotics in the near future and issues associated with their gradual accumulation in the environment. Moreover, the fact that allicin (in contrast to other antibiotics) is volatile means that the physical presence of the carrier in the vicinity of the target site, e.g., bacterial infection, is preferable but not critical. Therefore, volatile thiosulfinates can reach even regions not accessible for the typical drug carriers commonly utilized in topical and pulmonary delivery.

In this work, we investigated various carrier materials and optimized the process parameters of spray drying, e.g., the inlet temperature of the drying air, use of external cooling, and type of atomizer, with the goal of maintaining a high specific activity of encapsulated alliinase in the final product. Using the innovative non-contact diffusion method, we demonstrated that the spray-dried powders activated by air moisture or manually pre-wetted powder could be applied for *in-situ* generation of therapeutically effective doses of allicin. The knowledge about the relationship between carrier material and enzyme activity allows for prolonged allicin formation and avoids its high spatiotemporal concentrations or burst release, leading to undesired side effects such as skin irritation or severe chemical burns. These findings may contribute to the development of sustainable nature-inspired products based on the controlled *in-situ* synthesis of highly reactive antibiotics from their stabilized precursors.

1. Stokes, J.M., et al., A Deep Learning Approach to Antibiotic Discovery. Cell, 2020. 180(4): p. 688-702.e13.
2. Gupta, K. and R. Viswanathan, Combined Action of Streptomycin and Chloramphenicol with Plant Antibiotics against Tubercle Bacilli. Part I: Streptomycin and Chloramphenicol with Cepharanthine. Part II: Streptomycin and Allicin. Antibiotics & Chemotherapy, 1955. 5(1): p. 24-7.
3. Leontiev, R., et al., A comparison of the antibacterial and antifungal activities of thiosulfinate analogues of allicin. Scientific reports, 2018. 8(1): p. 1-19.
4. Bhattacharya, S., et al., Kinetics of bactericidal potency with synergistic combination of allicin and selected antibiotics. Journal of bioscience and bioengineering, 2022. 133(6): p. 567-578.

653b Simple Experimental Methods to Calibrate DEM Input Parameters for Pharmaceutical Tablet Applications

Tim Hoogenraad, Lucilla Coelho de Almeida, Edward Yost, Carl R. Wassgren, Ariel Muliadi

Tablet is the most common dosage form for small molecule drugs. Tablets are typically made by compressing a blended formulation containing some active pharmaceutical ingredients using a series of punch and die sets installed on a rotary press. After the tablets are made, they are often subjected to additional processing and handling that may damage them, such as dedusting, coating, packaging, and transportation/distribution. To ensure that formulated tablets can 'survive' the above processes, formulators often test the tablet's friability during the tablet development stage. The industry standard for such a test is summarized in USP <1216>. However, it is important to note that existing experimental tests, that described in USP <1216> included, may not fully represent the actual circumstances tablets face during manufacturing and further downstream processing.

To address such a shortcoming, the pharmaceutical industry has increasingly relied on alternate approaches to better simulate factors leading to tablet wear. In recent years, the progressive switch from empirical-based development to model-based quality-by-design as applied to pharmaceutics has triggered the proliferation of the use of predictive numerical methods such as the discrete element method (DEM). This simulation technique allows for the analysis of granular materials, including powders and tablets, by treating them as individual particles. By utilizing DEM, it becomes possible to virtually predict tablet dynamics during processing. However, researchers face a significant challenge in calibrating the tablet properties used in these models.

Model calibration is the initial step in any DEM modeling study and typically involves simulating an experimental system. The particle parameters in the model are then systematically varied until the predicted bulk quantities match the experimental counterpart. Since DEM simulations tend to be computationally intensive, the experimental system used for this calibration exercise must ideally satisfy the following criteria: (1) it can be easily set up experimentally and numerically, (2) it does not involve too many particles, and (3) the bulk behavior of the particles/tablets produced by the system is strongly influence by the particle parameter.

This work presents a series of simple experimental tests meeting the above criteria specifically for calibrating the DEM particle parameters relevant to simulating bulk tablets and wear behaviors in common pharmaceutical operations. Ansys Rocky software package was used for the DEM simulations. The DEM model incorporates real tablet shapes (instead of some glued-spheres representations). A hysteretic linear spring model and a linear spring Coulomb limit model were used to compute, respectively, normal contact and tangential forces.

We focus on the calibration of two groups of DEM parameters. The first group includes internal parameters such as particle density, Young's modulus, and Poisson's ratio. The second group comprises parameters that describe material interactions, which encompass coefficients of restitution and the static and dynamic friction coefficients between one particle and another as well as between the particles and the solids boundaries.

The calibration methods are covered in detail, including a thorough evaluation of the significance of each test for the DEM parameters and their comparison with the experimental data. The accuracy of the calibrated parameters is demonstrated through validation experiments. We highlight which input parameters are important for accurate simulations of pharmaceutical applications such as coating or packaging and identify parameters of lesser importance. Lastly, we describe the limitations of existing experimental methods for assessing tablet friability (e.g. the USP <1216> method) and propose alternative numerical approaches that can better predict tablet wear in a given pharmaceutical process.

653c *DEM Simulations and Experiments: Study of Dry Impregnation and Its Dependence on Particle Morphology: Spheres Cylinders, Trilobes and Quadrilobes*

Maria Tomassone

The impregnation of metal solutions onto porous catalyst supports is one of the key steps for preparing industrial metal-supported catalysts. In this process, metal salts or complexes are first dissolved in an aqueous solution. The volume of the metal solution applied is the same as or less than the total pore volume of the catalyst support. When conducting dry impregnation in a rotating vessel, metal solution is typically sprayed over a granular bed containing a porous catalyst support such as alumina (Al_2O_3) or silica (SiO_2). During spraying, capillary action draws the metal solution into the pores and metal complexes are adsorbed onto the high surface area support particles.

A problem may arise as excess liquid forms liquid bridges between particles, which causes wet cohesive forces and disrupts flow and further affects granular mixing processes. As a consequence, this can also adversely affect the homogeneous dispersion of the metal precursor in the particle.
Due to the large volumes and high cost associated with producing heterogeneous catalysts, several open questions regarding the optimization of this process remain: (i) how mixing and flow are affected when the particles have a certain degree of moisture or are saturated with fluid, (ii) how to improve fluid distribution to and within the granular bed (iii) the extent and distribution of dead zones for a given impregnator configuration, (iv) how does the morphology of the particles affect the uniformity of the liquid in the granular bed, (v) can the mixing and content uniformity be improved by using alternate geometries, such as baffles, spray pattern, or systems, (vi) how does the rotation pattern, nozzle and spray pattern, as well as the morphology of the catalyst support affect mixing profiles within the granular bed, and finally, (vii) what are the rules of thumb for process scale-up?

In this work, we performed DEM simulations to study impregnation in different porous support morphologies, spheres cylinders, trilobes, quadrilobes of different nominal diameters and aspect ratios.
We found that the time to achieve liquid uniformity in the particle bed for quadrilobes and trilobes of a given nominal diameter converges to the time for uniformity of a sphere with the same nominal diameter. (Aspect ratio (AR) equal to 1). Quadrilobes and Trilobes behave very similarly to cylinders of the same nominal diameter and for a given nominal diameter, the higher the AR the smaller the time to achieve uniformity. We observe that the time to achieve uniformity for particles, being cylinders, trilobes and quadrilobes with Aspect Ratio greater than 1 is always shorter than for spheres. We believe that higher probability of collisions correlates well with shorter times to uniformity. The length of the extrudate does not have a strong influence in the S/V ratio after a nominal diameter of 6mm. This ratio converges to a fixed number for a given diameter after 6mm, however, for shorter lengths it matters.
The effect of flow and operating efficacy in impregnation (adsorption of liquids, adsorption of chemicals) through any extrudate bed would be influenced not only by the Surface to volume (S/V) ratio but also by the packing fraction of the extrudate and the tap density of the material and both would be influenced by the length of the cut extrudate pieces
It is all these factors and not only S/V ratio that will determine the final operation functionality of the impregnation of the material in a bed. In addition, we performed 18 experiments with metal solutions, 3 supports and 2 different concentrations. Regardless of the type of support, titania or alumina, we observe that particles with larger aspect ratios always fall below spheres as predicted by simulations. Trilobes are much longer particles and more uniform aspect ratio. This is consistent with simulations:

particles of larger AR reach uniformity faster: Trilobes show better uniformity than cylinders. Also, for cylinders, we observe that larger cylinders show uniformity faster as expected since they have smaller S/V. The time to achieve uniformity follows the same trends predicted by simulations.

653d Early-Stage Materials Characterization for Predicting Late-Stage Tablet Breakage during Film-Coating

Jeffery Larson, Annika Lee, William Ketterhagen

Film coats are often applied to pharmaceutical tablets for a variety of aesthetic and/or functional purposes. Given many years of industrial experience, the tablet coating process is typically conducted successfully and without issue. Based on a previously discussed case of tablet breakage that led to a successful predictive approach [1], we generalized this approach to seek a better understanding of the role tablet properties and impact conditions have in tablet breakage. We found that observed tablet breakage probabilities are well characterized by the theoretical breakage model of Vogel and Peukert [2]. In this work, we focus material parameters (such as mass-based material strength (F_{mat}) [2], mass specific threshold energy ($W_{m,min}$) [2], peak impact fracture force [3], etc...) to develop a material/time sparing approach that enables early-stage tablet assessment for large scale coating pan success in late-stage development. Identification and determination of the approach for robust drug product development are discussed.

References

[1] Ketterhagen, W. R., Larson, J., Spence, K., & Baird, J. A. (2021). Predictive approach to understand and eliminate tablet breakage during film coating. *AAPS PharmSciTech*, 22(5), 178.

[2] Vogel, L., & Peukert, W. (2003). Breakage behaviour of different materials—construction of a mastercurve for the breakage probability. *Powder Technology*, 129(1-3), 101-110.

[3] Alhusban, F., & Murgatroyd, E. F. (2024). Impact testing as a new approach to determine mechanical strength of pharmaceutical tablets. *International Journal of Pharmaceutics*, 123891.

Disclosures

AbbVie contributed to the design; participated in collection, analysis, and interpretation of data; and in writing, reviewing, and approval of the final version. AbbVie sponsored and funded the study. JL and WK are current employees of AbbVie and may own AbbVie stock. AL previously interned at AbbVie and has no conflict of interest.

653e Assessment of Air Entrapment Related Tablet Defect Risk in Tablet Manufacturing

Yaser Bozorgi, Subhash Thakur, Jeffrey Wang, Pingjun Tang

Excessive entrapped air during high-speed powder compaction poses risk in pharmaceutical tablet production, often leading to observable defects such as fissures and lamination. This study investigates the relationship between air pressure, tensile strength, and the risk of air entrapment in tablets using a 1D transient air entrapment model developed by Zavaliangos et al. (2017). The model is applied to the compression of various active and excipient blends to analyze the resulting tablet air pressure profiles.

To facilitate the model's application, tablets varying in density were made using a compaction simulator. The permeability and tensile strength of the tablets were measured as functions of relative density. Permeability measurement at low-to-moderate tablet relative densityconducted using a FT4 powder rheometer, and at high tablet relative density were performed using a Pulse Permeameter. Additionally, the tensile strength of each tablet was measured using a diametrical tensile strength tester.

The simulation results align with observations, indicating that formulations with a high air pressure-tensile strength ratio exhibit signs of excessive air entrapment related tablet defects. This research highlights the potential of utilizing this hybrid modeling and experimental approach as a material-sparing methodology for assessing air entrapment risks during the early stages of formulation development.

653f M^2E^3d: Evolutionary Equation Discovery and Its Applications in the Powder-Handling Industries

Andrei-Leonard Nicusan, Owen Jones-Salkey, Kit Windows-Yule

Multiphase systems are vital to the development of a myriad of products spanning multiple industrial sectors, from the formulation of novel biofuels, chemicals and pharmaceuticals in stirred-tank reactors to plastic recycling in gas-fluidised beds. Numerical models of these and many other systems are used to better understand and optimise their internal dynamics and end-products. Modelling classical fluids is relatively simple: their properties may be directly measured using widely-available characterisation tools, and directly input to CFD models. Particulate media - powders, granulates - for which there exist no established continuum models, are significantly harder to simulate; they are typically modelled piece-wise using the Discrete Element Method (DEM), whose calibration is infamously problematic . Modelling multiphase systems is more complex still, requiring detailed properties of both fluid and particulate phases, and suitable drag models for coupling. There exists a dizzying array of (often imprecise) empirical models for this task (Luding S. Introduction to discrete element methods: basic of contact force models and how to perform the micro-macro transition to continuum theory. European journal of environmental and civil engineering. 2008).

Multiphase Materials Exploration via Evolutionary Equation Discovery M^2E^3D) is a novel evolutionary approach to determine the equations governing their interactions. In other words, M^2E^3D **autonomously discovers the equations required to accurately reproduce a system's full, three-dimensional dynamics**. The resulting combinatorial optimisation problem of discovering analytical closed-form expressions modelling previously-measured data points is solved using a symbolic regression engine employing an evolutionary algorithm balancing the model's error and complexity.

M^2E^3D has been rigorously validated, using the technique to autonomously (re)discover well-known analytical laws such as Stokes drag on spherical particles and Darcy-Weisbach pressure drop in pipes using only 32 measurements. It has then been applied to derive entirely novel relations for complex industrial systems that are impossible to model analytically, such as:

- Predictive models for powder blending performance in continuous direct compression (paper in collaboration with AstraZeneca; Jones-Salkey O, Windows-Yule CR, Ingram A, Stahler L, Nicusan AL, Clifford S, de Juan LM, Reynolds GK. Using AI/ML to predict blending performance and process sensitivity for Continuous Direct Compression (CDC). International Journal of Pharmaceutics. 2024).
- Correlations predicting pharmaceutical tablet tensile strength from process conditions, with applications in online process control (paper in collaboration with GlaxoSmithKline, under review).
- Scaling laws for 3-phase fluidised beds as functions of liquid loading and viscosity.

In this talk we introduce the methodology underlying M^2E^3D and demonstrate its application through a series of case studies. The Julia, Python and C++ algorithms developed are open-source, portable and massively-parallel, and were used successfully from laptop-scale shared-memory machines to multi-node supercomputing clusters - facilitating high-throughput discovery and understanding of systems beyond those explored here.

571b Optimal Control of Batch Crystallization Processes: New Methods for New Challenges

Hao-Jen Pan, Wei-Ting Chien, Jeffrey Ward

Batch crystallization is a vital technique for the separation and purification of substances, especially in the production of pharmaceutical and fine/specialty chemicals. The operation of such processes can be optimized to improve crystal properties such as size and shape distribution. In general, to model a batch crystallization process it is necessary to solve a partial differential equation to determine how the crystal size distribution changes with time during the batch. If the process must be simulated repeatedly for optimization, this may be time consuming if a high resolution is desired. Furthermore, at the end of the batch the exact crystal size distribution is not known, only the moments or the values of the CSD at certain discrete points.

Raisch and coworkers [1-5] demonstrated that after a specific transformation of variables, optimal control theory can be applied to solve certain optimization problems in batch crystallization very efficiently and nearly-analytically, recovering the complete crystal size distribution. In the present work their method is modified, extended and applied to a variety of other problems of practical interest.

First, the method is applied to determine optimal operating trajectories for a large set of crystallization systems using a dimensionless framework [6]. The framework is used to calculate Pareto fronts for the tradeoff between the number of nuclei and the nucleated mass. The results show that for most systems operation with constant supersaturation gives a result similar to that of the knee point on the Pareto front suggesting that the constant supersatuation trajectory (which can be determined without a crystallization kinetic model) is a reasonable choice when the crystallization kinetics are unknown.

Second, the method is applied to systems with primary nucleation [7]. The original work of Raisch and coworkers considered only secondary nucleation where the nucleation rate is proportional to the growth rate and the magma density raised to certain powers. In the present work it is shown that the previous method can be extended to consider primary nucleation. The resulting two-point boundary value problem is difficult to solve using conventional shooting methods because of the equations are highly nonlinear. Instead the boundary value problem is solved using a gradient method. The algorithm is found to be fast, accurate and robust and permits the study of a tradeoff between number of nuclei and nucleated mass.

Third, the method is applied to batch cooling crystallization processes where the crystallization kinetics are temperature-dependent [8]. Three strategies for this case are considered and compared. The first strategy is to neglect the nucleated mass when calculating the nucleation rate as first suggested by Hofmann and Raisch [2]. The second strategy is to solve the two-point boundary value problem using a gradient method. The third strategy is to combine the two, using the approximation to determine an initial guess for the gradient method. The results show that the approximation introduces little error and the computational burden of the gradient method is dramatically reduced if the result from the approximation is used as an initial guess.

Fourth, the method is applied to problems with nucleation and crystal shape [9]. Pareto fronts for the tradeoff between number of nuclei and nucleated mass are plotted subject to constraints on the product crystal shape. The endpoints of the Pareto fronts map out a feasible region for the operation of the process, and the tradeoff between objective functions is most pronounced near the midpoint between the minimum and maximum achievable aspect ratio of the seed-grown crystals.

Finally, the method is applied to problems where cycles of growth and dissolution are used to tailor the product crystal size distribution [10]. In this case, the method can be used to efficiently determine the attainable region (in terms of seed size and seed variance). The results show that the attainable region expands as the number of cycles of growth and dissolution is increased, but the extent of the change of the crystal size depends on the sensitivity of the growth and dissolution rates to the crystal size.

References

1. Bajcinca, N.; Hofman, S. Optimal control for batch crystallization with size-dependent growth kinetics. In *Proceedings of the 2011 American Control Conference*, **2011**; IEEE: pp 2558-2565.

2. Hofmann, S.; Raisch, J. Application of optimal control theory to a batch crystallizer using orbital flatness. In *16th Nordic Process Control Workshop, Lund, Sweden*, **2010**; pp 25-27.

3. Angelov, I.; Raisch, J.; Elsner, M. P.; Eidel-Morgenstern, A. S. Optimal operation of enantioseparation by batch-wise preferential crystallization. *Chem. Eng. Sci.* **2008**, *63* (5), 1282-1292. DOI: 10.1016/j.ces.2007.07.023.

4. Vollmer, U.; Raisch, J. Control of batch crystallization - A system inversion approach. *Chemical Engineering and Processing-Process Intensification* **2006**, *45* (10), 874-885. DOI: 10.1016/j.cep.2006.01.012.

5. Vollmer, U.; Raisch, J. Control of batch cooling crystallization processes based on orbital flatness. *Int J Control* **2003**, *76* (16), 1635-1643. DOI: 10.1080/00207170310001626419.

6. Pan, H. J.; Ward, J. D. Dimensionless Framework for Seed Recipe Design and Optimal Control of Batch Crystallization. *Ind Eng Chem Res* **2021**, *60* (7), 3013-3026. DOI: 10.1021/acs.iecr.0c06132.

7. Chien, W. T.; Pan, H. J.; Ward, J. D. Optimal Control of Batch Crystallization Processes with Primary Nucleation. *Ind Eng Chem Res* **2023**, *62* (22), 8834-8846. DOI: 10.1021/acs.iecr.3c00286.

8. Pan, H. J.; Ward, J. D. Algorithms for solving boundary value problems in optimal control of seeded batch crystallization processes with temperature-dependent kinetics. *Chem. Eng. Sci.* **2023**, *270*. DOI: 1016/j.ces.2023.118517.

9. Pan, H. J.; Ward, J. D. Optimization of Simple Batch Crystallization Systems Considering Crystal Shape and Nucleation. *Ind Eng Chem Res* **2020**, *59* (20), 9550-9561. DOI: 10.1021/acs.iecr.9b06842.

10. Pan, H.-J.; Ward, J. D. Computationally Efficient Algorithm for Solving Population Balances with Size-Dependent Growth, Nucleation, and Growth-Dissolution Cycles. *Ind Eng Chem Res* **2021**, *60* (34), 12614-12628. DOI: 10.1021/acs.iecr.1c01947.

571c Polymer-Induced Aggregation of Hematite Under Shear By CFD-DEM Simulations

Lequan Zeng, George Franks, Eirini Goudeli

Effective particle recovery by froth flotation predominantly occurs within particle size ranges of 50-150 microns [1]. As such, aggregation of fine valuable particles smaller than 20 microns plays an important role for their recovery during mineral processing. Mineral microparticles can spontaneously undergo weak aggregation under shear due to attractive interparticle (DLVO) interactions [2]. However, aggregates smaller than 50 microns are not sufficiently large to be effectively separated via froth flotation. Addition of polymers to the suspension is commonly used to produce large and strong mineral aggregates [3], that are suitable for collection via froth flotation. The mechanism of polymer-induced aggregation, as well as the detailed relationship between particle interactions and polymer dosage, and the dynamics of aggregate growth remain poorly understood. Particle-based simulations offer a promising approach for investigating strong aggregation induced by polymers, enabling the quantification and observation associated with the aggregation process.

Here, coupled Computational Fluid Dynamics – Discrete Element Method (CFD-DEM) is employed to explore the impact of polymer dosage on size and structure of hematite particle under shear, taking into account both particle-particle and particle-fluid interactions. The Yukawa force model [4, 5], is introduced to describe polymer-induced bridging flocculation for strong particle aggregation. The temporal evolution of aggregate size, size distribution, number concentration, and structural characteristics during aggregate growth are quantified shedding light into the detailed mechanism of polymer-induced aggregation.

References

1. Ralston, J., Fornasiero, D., Grano, S., Duan, J., and Akroyd, T., *Reducing uncertainty in mineral flotation—flotation rate constant prediction for particles in an operating plant ore.* International Journal of Mineral Processing, 2007. **84**(1): p. 89-98. https://doi.org/10.1016/j.minpro.2006.08.010.

2. Zeng, L., Franks, G.V., and Goudeli, E., *Aggregation and breakage dynamics of alumina particles under shear by coupled Computational Fluid Dynamics – Discrete Element Method.* Journal of Colloid and Interface Science, 2024. **661**: p. 750-760. https://doi.org/10.1016/j.jcis.2024.01.210.

3. Zhou, Y., Gan, Y., Wanless, E.J., Jameson, G.J., and Franks, G.V., *Interaction Forces between Silica Surfaces in Aqueous Solutions of Cationic Polymeric Flocculants: Effect of Polymer Charge.* Langmuir, 2008. **24**(19): p. 10920–10928. https://doi.org/10.1021/la801109n.

4. Elghazrani, K., Azougarh, A., Oberdisse, J., and Filali, M., *Interactions between microemulsion droplets decorated with hydrophobically modified polymers: A small-angle neutron scattering study.* The European Physical Journal E, 2014. **37**(12): p. 128. https://doi.org/10.1140/epje/i2014-14128-8.

5. Yukawa, H., *On the Interaction of Elementary Particles. I.* Proceedings of the Physico-Mathematical Society of Japan. 3rd Series, 1935. **17**: p. 48-57. https://doi.org/10.11429/ppmsj1919.17.0_48.

571d Investigating the Effect of Temperature on Calcium Sulfate Scale Formation Using a Novel Continuous Stirred Tank Reactor (CSTR) Laser Setup.

Glavic Tikeri, Ali Alshami

In the oilfield industries, the rate and extent of calcium sulfate scale formation, a prevalent issue affecting operational efficiency and costs, varies significantly with temperature. My research involves investigating the extent and dynamics of this variation using a novel continuous stirred tank reactor (CSTR) setup integrated with laser techniques. Calcium sulfate is one of the most problematic and hardest scales posing substantial challenges across diverse industrial sectors, including water treatment, oil production, and mineral processing. To address these challenges, a CSTR system is implored to replicate scale formation under meticulously controlled temperature conditions. Concurrently, laser techniques are harnessed for real-time monitoring and characterization of the scaling process, offering unprecedented insights into its dynamics. Using a temperature range of 0 to 100°C within the CSTR, the experiment scrutinizes the influence of temperature on both the rate and magnitude of scale formation. My results underscore a pronounced escalation in scale accumulation alongside a notable reduction in induction time as temperature increases, from about 4 hours at room temperature to just 4 minutes at 100°C. Furthermore, scanning electron microscopy (SEM) analysis was done on the collected scale crystal samples and temperature-induced alterations in crystal morphology were noticed revealing significant crystal degradation with increase in temperature. These findings furnish indispensable insights for optimizing process parameters and devising robust scale mitigation strategies in various industrial applications. By comprehensively understanding the interplay between temperature and calcium sulfate scale formation, industries can proactively mitigate scale-related challenges, thereby enhancing operational efficiency and minimizing costs.

571e Custom Fluidized Bed Reactor for Atomic Layer Deposition on Li-Ion Battery Cathode Powder

Julie A. Nguyen, Michael Zaza, Kent J. Warren, Alan W. Weimer

The first fluidized bed reactor for atomic layer deposition (ALD) on particles was developed almost two decades ago, which helped the ALD technology expand beyond the semiconductor industry to catalysis and energy storage. In particular, particle ALD can be applied to energy storage materials like nickel-rich cathode active materials like $LiNi_xMn_yCo_{1-x-y}O_2$ (NMC) to stabilize the surface and improve electrochemical performance for Li-ion battery applications. With sufficient cycles the ALD process deposits conformal, pin-hole free films with Angstrom-level precision. However, nickel-rich NMC such as $LiNi_{0.8}Mn_{0.1}Co_{0.1}O_2$ (NMC811) is an air-sensitive material that requires handling under inert conditions to prevent growth of surface contaminants like hydroxides and carbonates. Here, we show a unique fluidized bed reactor design that allows inert (un)loading of material and real-time quantification of vapor species evolved from the ALD reactions. NMC811 was coated using the trimethylaluminum (TMA)/water chemistry. The fluidization behavior of NMC811 powder was characterized and the minimum fluidization velocity calculated. Furthermore, the quality of the film deposited by the system was assessed using inductively coupled plasma optical emission spectroscopy (ICP-OES), Brunauer-Emmett-Teller (BET) analysis, and low energy ion scattering (LEIS). Controlled delivery of calibration gas to an on-line mass spectrometer allowed subsequent quantification gaseous species evolving from the NMC811 powder during the TMA and water doses. Our present work showcases a fluidized bed reactor capable of performing ALD on air-sensitive particles, in addition to quantifying the vapor species evolving from the powder. This reactor design can help inform construction of future fluidized bed reactors that coat air-sensitive material.

571f Production of Catalyst Tablets: Effect of Powder Properties

Mingzhi Zhao, Sai Yellapragada, Christine Hrenya, Benjamin Glasser, Alberto Cuitino

Compacted aluminum oxide, or 'Alumina' tablets, are commonly used as catalyst support to facilitate efficient chemical processes. We investigated the effect of moisture content on various types of alumina tablets and found that elevating the moisture content of the granulated alumina up to 20% w/w will increase the mechanical properties of the tablets, such as hardness, Young's modulus, and tensile strength. Furthermore, heating the tablets will alter these mechanical properties.

571g Evaluation of Microcrystalline Cellulose Powder Flowability with Varying Particle Properties and Moisture Content

Jordan Monroe, Heather Emady

Understanding particulate flow is crucial to the design and development of solids handling in industries from foods to pharmaceuticals. The relationship between particle properties (micro-scale), bulk particle flow (meso-scale), and for a broad range of powders is not well established due to the complexity of solids systems.

This work experimentally investigates the effects of particle size and moisture content on the flowability of micro-crystalline cellulose (MCC). The raw MCC was sieved into varying size fractions based on the mesh size ranges of 0-25 µm, 25-45 µm, 45-63 µm, 63-90 µm, and 0-90 µm. Particles in each size fraction were characterized by optical microscopy via a Malvern Morophologi G3SE Particle Analyzer for their and shape descriptors. Using a Freeman FT4 Powder Rheometer, bulk powder properties to characterize flowability were found, including basic flowable energy, compressibility, flow function coefficient, etc. Moisture content was varied via the addition of deionized water in amounts from 0% to 50% by total weight. Particle size, shape, and moisture content were correlated with the FT4 flowability results. This is the first study, to our knowledge, to quantify the effects of particle shape, size, and moisture content on the flowability of powders.

571h Heat Transfer in a Rotary Drum: Effect of Equipment Design

Carlin Leung, Marcella Alves, Elaheh Ardalani, William Borghard, Nina C. Shapley, Alberto Cuitino, Benjamin Glasser

Rotary calciners and dryers are a common equipment used in a number of industries such as in pharmaceutical, catalysts, food industry, and chemical industry. Calciners and dryers are common equipment used for thermal treatment for several materials. Understanding heat transfer and particle flow can reduce wasted energy and materials and significantly improve the product quality. For good product quality, it is a necessity to uniformly increase the material temperature. Typically, rotary drums are made of steel, but can be made of ceramic or cast iron, each having a different surface roughness and texture. Surface roughness can have a significant effect on heat transfer and temperature uniformity on the material. Studies have been done on mixing of materials in rotary calciners and dryers. However, to the best of the authors' knowledge there are no studies focusing on heat transfer based on equipment surface roughness in a rotary drum.

Understanding the relationship of equipment surface roughness on heating time is important for predicting processing time in real-world applications. In this study, simulations using the discrete element method (DEM) and the Hertz-Mindlin contact model were used to better understand the effect of the coefficient of static friction (surface roughness) on heat transfer for non-cohesive particles. The model accounts for particle-to-particle and particle-to-wall interactions. The surface roughness between particle and wall and the number of baffles in the rotary drum were studied. We saw that static friction has a large effect on thermal time and temperature uniformity. Depending on the static friction tested, two flow regimes were exhibited: Slipping and rolling regime. At low static frictions, thermal time is high and temperature uniformity is poor, while increasing static friction, decreases thermal time and improves temperature uniformity. Total kinetic energy of the particles and particle tracking were investigated to further understand the effect of surface roughness on flow patterns and heat transfer. Baffles were added in the simulations to investigate their effect on heat transfer for different values of static friction. We found that adding baffles for low static friction cases greatly improves heat transfer and temperature uniformity, but consequently increases the total kinetic energy of the particles, which can lead to attrition. This work provides quantitative findings on how equipment design affects heat transfer and temperature uniformity in a rotary drum.

571i Innovative Design and Performance Assessment of a Novel Modular Reactor for One-Step Liquid Fuel Production from Stranded Natural Gas with Co-Utilization of Carbon Dioxide.

Tanay Jawdekar, Sudeshna Gun, Congwen Lu, Anuj Joshi, Sonu Kumar, Joel Paulson, Liang-Shih Fan

With increasing global energy demands, the market for fossil-based fuels is projected to grow substantially over the next 30 years. Oil and gas, being the major fossil-based energy sources for the transportation and industrial sector, have demonstrated an increased crude extraction by ~120% in the past decade. However, the crude extraction process faces a challenge in tackling the associated natural gas that is dissolved in the crude oil or prevails in the free gas caps of oil reservoirs. Considering the risks and costs involved in natural gas transportation, the quantity of associated gas is very small to substantiate the construction of a pipeline network. Although this associated natural gas is the by-product of the oil extraction process, the absence of a viable solution for utilization makes it 'stranded natural gas (SNG).' Approximately 15 tons/day of SNG is flared from one oil well, contributing to greenhouse gases. The uneconomical nature of pipeline installation to transport SNG and the necessity to find a robust solution to tackle SNG opens an avenue for developing modular technologies that can convert SNG to value-added products while reducing CO_2 emissions. Liquid fuels (LF) are high-density energy carriers that are easy to transport and can support the existing infrastructure for energy consumption. 15 tons/day of SNG that is flared has the potential to produce ~2.5 tons/day of LF and generate ~115 kW of electricity simultaneously. This work proposes a single reactor modular unit that converts SNG to value-added LF. The novel reactor configuration consists of a multi-tubular packed bed setup divided into three intermediate sections: a mixed reforming section, a section for heat exchange, and a Fischer-Tropsch (FT) section. Mixed reforming of SNG uses CO_2 and steam to produce high-quality syngas (H_2:CO = 1.7), which in turn can be used to produce liquid fuels in the FT section of the reactor. The heat exchange section assists in increasing the efficiency of the process by facilitating heat integration within the process streams. This process provides inherent CO_2 utilization and converts pressurized SNG to easy-to-transport liquid fuels that can be processed to manufacture motor spirit, diesel, jet engine fuels, and feedstocks for the petrochemical industry. System-level thermodynamic evaluation was conducted using ASPEN Plus software for the production of 2.5 tons/day of liquid fuel by taking into consideration multiple plausible cases of reactor configuration and heat integration. Detailed economic calculations performed based on the tubular reactor design for all configurations indicate that the production of liquid fuels using SNG can prove to be a profitable venture. The design of the multi-tubular reactor and the economics associated with it are simultaneously optimized using the Bayesian optimization technique to propose a portable, modular reactor requiring a minimum number of units to process the SNG from an oil well. The robustness of the best-case scenario is then validated for fluctuating SNG compositions, flow rates, and temperatures. This work establishes the design feasibility of a novel, economically viable, modular technology for upgrading SNG to value-added liquid fuels while reducing greenhouse gas emissions.

571j Highly Porous Coatings of Metal Nitrides on Electronic Circuitry for Sensing Applications

Adrien Baut, Andreas Güntner

Molecular sensing is important for medical diagnostics, environmental monitoring and food safety [1]. Metal nitrides are an emerging class of materials that feature promising surface properties for the chemoresistive sensing of gases at low temperature. Yet, the template-free deposition of thin, porous metal nitride coatings is a challenge, motivating the design of novel processes. Here, we will present a new pathway to design highly porous (>90%) metal nitrides coatings on electronic circuitry by combining aerosol deposition and a subsequent dry conversion [2]. We will demonstrate close control over process parameters that affect film morphology, as monitored *in situ* by measuring the coatings resistance. The obtained coatings is rather homogeneous and consist of a network of fine and highly crystalline nanoparticles (crystal size ~15 nm), as revealed by X-ray diffraction and Scanning Electron Microscopy. To demonstrate immediate practical impact, we will show the selective sensing of NO_2, a major pollutant, down to few parts-per-billion concentration at room temperature.

References:

1. Güntner, A.T.; Pineau, N. J. & Pratsinis, S.E., *Energ. Combust. Sci.* **2022**, *90*, 100992.
2. Güntner, A.T.; Wied, M.; Pineau, N.J.; Pratsinis, S.E., *Sci.* **2020**, *7* (7), 1903390.

571k CO_2 Free Hydrogen Production Via Methane Pyrolysis Using a Moving Packed Bed Approach

Shekhar Shinde, Danwyn Aranha, Ishani Karki Kudva, Rushikesh Joshi, Krutarth Pandit, Liang-Shih Fan

The contemporary surge in global energy demands exerts escalating pressure on fossil fuel resources, predominantly due to the consequential generation of carbon dioxide (CO_2) and its exacerbation of the greenhouse effect. Consequently, scientists worldwide are earnestly exploring alternative energy sources characterized by reduced carbon intensity relative to conventional fossil-based energy. Diverse methodologies have been devised for the synthesis of fuels, including methanol, syngas, and hydrogen. Of particular interest within the scientific community is hydrogen, owing to its emission-free combustion properties and high calorific values.

Extensive research has been undertaken on hydrogen production, encompassing electrochemical, thermochemical, and biological pathways. This investigation is specifically centered on the thermochemical process of methane pyrolysis for hydrogen production.

Methane pyrolysis entails the thermal cleavage of methane molecules into hydrogen and carbon. During methane pyrolysis, the thermal splitting of methane occurs within a reactor, where the methane molecules are subjected to elevated temperatures, typically exceeding 1000 °C. The resulting hydrogen is liberated in a gaseous state, while solid carbon is deposited on a catalyst material. Diligent efforts have been devoted to harnessing the deposited carbon through the catalytic generation of Carbon Nanotubes (CNTs), graphitic carbon, etc. The removal of this carbon is predominantly executed through a dislodging mechanism facilitated by a fluidized bed.

Currently, fluidized bed strategies are predominantly used in both methane pyrolysis and carbon dislodging. Our endeavor is to introduce a moving bed strategy for methane pyrolysis, enhancing control over reaction parameters to achieve heightened methane conversions and superior hydrogen purity. The current research focuses on elucidating the kinetics of the Methane Pyrolysis technology and its integration into a Technology Readiness Level (TRL) 5 process. The advancement pertaining to the dislodging of deposited carbon will be demonstrated at the TRL 5 level, employing two distinct strategies: one utilizing a moving bed system with lower gas velocities and the other operating in the pneumatic conveying regime of a fluidized bed system with higher gas velocities. Each strategy involves a unique gas-solid fluidization concept, differentiating the system from conventional approaches.

The resulting hydrogen exhibits high purity, rendering it suitable for diverse applications. Notably, due to its capacity to produce top-tier H_2 without CO_2 emissions, the integration of our process with carbon-intensive applications such as ammonia production and the steel industry holds the potential to mitigate the carbon intensity associated with these processes significantly.

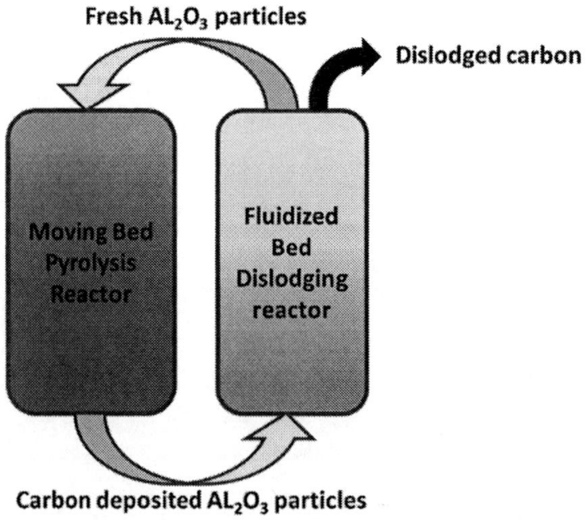

571l Novel Sorbent Particle for CO2 Capture and Utilization

Seyedamin Razavi, Vahid Rahmanian, Runxia Cai, Mahe Rukh, Mohammadreza Kosari, Saad A. Khan, Fanxing Li

This study reports two distinct families of sorbents for CO_2 capture and combined capture and utilization. In the first part of the study, we report perovskite oxides as a new family of tunable sorbents. Perovskite oxides offer excellent flexibility to tailor the sorption kinetics and thermodynamic parameters. Using $Sr_xLa_{1-x}FeO_3$ as a basis, we showed that changing the A-site cation composition is highly effective in adjusting the CO_2 adsorption and desorption characteristics, allowing tunable CO_2 capture and release. A strong correlation between the oxide surface area and sorption capacity was also established. Despite the low surface area inherent to perovskite oxides, we enhanced their surface area from <5 m^2/g to ~30 m^2/g using a electrospinning method. $Sr_{0.2}La_{0.8}FeO_3$ prepared via electrospinning exhibited a relatively moderate desorption temperature (~240°C) and a CO_2 sorption capacity of 0.68 wt.%. We also showed that the sorption material can be effectively utilized for CO_2 capture and methanation in a temperature range of 250 to 300°C.

In the second part of this study, we report hybrid organic-inorganic sorbent particles by functionalizing mesoporous SiO_2 spheres (MSS) with 3-Aminopropyl triethoxysilane (APTMS) to introduce amine functional groups. In this study, MSS-(x)NH_2 (where x= 0.1, 0.5, and 1) was synthesized for CO_2 adsorption, with x representing the volume of APTMS per gram of MSS. The adsorption capacity of these samples was assessed using TGA, revealing that MSS-(1)NH_2 exhibited the highest capacity, with a 2.5% CO_2 sorption capacity and excellent stability.

571m Particle Molecular Layer Deposition of Amine Films for CO_2 Capture Materials

Hailey Loehde-Woolard, Bergen Evans, Michael Zaza, Katarina Odak, Kent Warren, Alan W. Weimer

Particle Molecular Layer Deposition (MLD), closely related to Atomic Layer Deposition (ALD), involves a series of alternating gas phase reactions on a particle surface. Our work studies particle MLD as a method of generating aminosilane films on a variety of solid supports to create CO_2 capture materials. In addition, we describe the construction and validation the custom particle ALD/MLD system used in this work.

In the past few decades, solid adsorbent materials have been a focus of research as an alternative to traditional liquid amine absorbent systems which require substantial regeneration energy and cost. Supported amine adsorbents have gained much interest as it maintains the traditional acid-base chemistry, operates under ambient conditions, and is tolerant of humidity. However, the current liquid-based functionalization techniques have limitations; grafted materials tend to have lower adsorption capacities and impregnated materials tend to have poorer regeneration abilities. (Unveren, 2017) In addition, liquid phase grafting is heavily dependent on numerous process conditions and produces amorphous 3-dimensional films. (Yadav 2013) Vapor deposition of aminosilanes has been demonstrated to produce reliable mono- and sub-monolayer films, which provide greater surface control but has not been studied in depth for CO_2 capture applications. (Yadav 2013, Rozyyev 2021) Vapor deposition also differs from MLD, as vapor deposition only exposes the surface to one reactant whereas MLD regenerates reactive groups on the surface through a series of reactions in each cycle.

Here, we use MLD to deposit a covalently bonded aminopropylsiloxane network on particle supports. A custom vibrating fluidized bed particle ALD reactor was constructed in house and validated through fluidization experiments and deposition of aluminum oxide as a benchmark chemistry. Following validation, the system was employed to investigate the MLD chemistry of (3-aminopropyl)triethoxysilane (mono-amine) and water. Sorbent materials were generated according to design of experiments that focused on factors related to film formation such as the number of cycles and deposition temperature. For comparison, materials were also generated through traditional liquid phase grafting and vapor phase deposition. Characterization of the sorbents included BET surface area analysis, LECO elemental analysis, X-ray photoelectron spectroscopy (XPS), and thermogravimetric analysis (TGA). The adsorption capacity was shown to be affected by both the number of MLD cycles and the temperature of deposition. It was previously found by Rozyyev et al. that for vapor deposition, lower deposition temperature led to greater amine content. Interestingly, MLD materials produced at higher temperatures lead to higher adsorption capacity.

Ünveren, E. E.; Monkul, B. Ö.; Sarıoğlan, Ş.; Karademir, N.; Alper, E. Solid amine sorbents for CO 2 capture by chemical adsorption: A review. Petroleum 2017, 3 (1), 37

Weimer, A.W. Particle atomic layer deposition. J. Nanopart. Res. 2019, 21:9

Yadav, A. R., Sriram, R., Carter, J. A., Miller, B. L. Comparative study of solution-phase and vapor-phase deposition of aminosilanes on silicon dioxide surfaces. Materials Science and Engineering C, 2014, 35, 283-290

Rozyyev, V., Murphy, J. G., Barry, E., Mane, A. U., Sibener, S. J., Elam, J.W. Vapor-phase grafting of a model aminosilane compound to Al_2O_3, ZnO, and TiO_2 surfaces prepared by atomic layer deposition. Applied Surface Science, 2021. 562, 149996

571n $CaCO_3$ Crystallization in the Presence of Polycarboxylated CMC: Insights into Kinetics, Mechanisms, and Crystal Morphology

Nadhem Ismail, Ali Alshami

Calcium carbonate crystallization is of interest due to its relevance in different applications such as pharmaceutical, biomineralization processes, catalysis and environmental systems. It is also used as an experimental system model to study the phenomena of crystallization and nucleation in general. Compared to additive-free systems, additive-controlled crystallization of calcium carbonate is still insufficiently explained due to the complexity and varieties of the possible interactions between the different species in the system involving crystals and additives. One of the interesting phenomena of additive-controlled systems is the threshold inhibition of nucleation and crystallization of calcium carbonate by polycarboxylates at very low concentrations (i.e. 1-20 ppm) in aqueous solutions. Additionally, a spontaneous and naturally occurring phenomena is observed through the biological organism's use of modified polysaccharides in stabilizing calcium carbonate. Understanding the influence of polycarboxylates on the calcium carbonate crystallization process is not only important in mineral scale inhibition but also crucial in elucidating biomineralization mechanisms where biomineral formation associated proteins are loaded with carboxylates (i.e. Aspartate residues).

This study explored, the impact of polycarboxylated carboxymethylcellulose (CMC) on the crystallization process of calcium carbonate. CMC was modified in a free radical polymerization reaction by grafting polyacrylic acid into the polysaccharide backbone, resulting in CMC-g-PAA. Different reaction conditions were explored to optimize a grafting copolymer with desired physicochemical properties. The crystallization process was investigated in a CSTR setup equipped with a laser system that is able to correlate crystal formation induction time with voltage difference. The real-time crystallization process done at different dosages of CMC-g-PAA is used to provide an information on the rate of the crystallization using the empirical Avrami equation. Results showed that CMC-g-PAA significantly affected the crystallization process even at low concentrations of 20 ppm. The impact of CMC-g-PAA on the crystallization was also analyzed by SEM and XRD. SEM images showed a big difference in the shape of the CMC-g-PAA treated crystals to the untreated crystals; long cylindrical crystals formed in the neat, additive-free, system compared to flowery shapes in the treated additive-controlled system. The exact phases and morphologies formed at different additive concentration levels are to be ascertained by analyzing XRD results. This study may result in better understanding of biomineralization processes and help combat undesirable calcium carbonate crystallization in different water-based environments.

571o Waste Valorization: High-Purity Syngas Generation from Co-Gasification of Waste Plastics and Biomass Via Chemical Looping Process

Ishani Karki Kudva, Shekhar Shinde, Tanay Jawdekar, Sudeshna Gun, Rushikesh Joshi, Sonu Kumar, Ashin Sunny, Liang-Shih Fan

The co-gasification of plastics and biomass is a process that combines the two feedstocks to generate several benefits. When plastics and biomass are co-fed into the gasifier, the high H/C ratio of plastics helps counterbalance the higher oxygen concentration in biomass, resulting in increased hydrogen yield, reduced tar generation, and improved syngas quality. The use of biomass, which is renewable and carbon-neutral, complements the energy content of plastics in co-gasification, resulting in a higher calorific value of the feedstock mixture and greater energy yields per unit of input. This process supports sustainability goals by decreasing dependence on fossil fuels and mitigating greenhouse gas emissions.

One of the most promising technologies for co-gasification is the chemical looping process, a two-reactor system that uses iron-based oxygen carriers to generate high-purity syngas. The feedstock is introduced into the first reactor, a moving bed reducer, where the iron-based oxygen carriers move concurrently with the feedstock. Steam is added along with feedstock for char gasification, and the oxygen carriers donate their lattice oxygen to carbon from the feedstock during this process, forming carbon monoxide (CO) and hydrogen (H_2). The result is high-purity syngas, i.e., CO+H_2, which can be upgraded to valuable chemicals in downstream processing. The second reactor, a fluidized bed combustor, regenerates the oxygen carriers using air and recirculates them back to the reducer. This process eliminates the need for expensive units like tar reformers, air separation, and acid gas removers.

The gasification process is highly endothermic, requiring a lot of energy. Several approaches have been explored to counteract this and achieve an autothermal system, i.e., no external heat is required. One of these methods involves burning biomass in the combustor to supply heat to the process, while another involves co-feeding natural gas, waste plastics, and biomass into the system. This serves two purposes: creating an autothermal system due to the high heating value of natural gas and using natural gas as a hydrogen source to produce higher-quality syngas. In the proposed strategies, the syngas purity is higher than 90%, and H_2:CO is approximately 2, which is required for downstream processing of valuable chemicals such as methanol.

Compared to the existing indirect gasification of mixed plastic waste, the proposed strategies outperformed them. The indirect gasification process relies on oxygen from an air separation unit to burn natural gas, generating heat for plastic gasification while producing carbon emissions, either treated by a highly energy-intensive acid gas removal process or left untreated. Therefore, the proposed strategies that use co-gasification of plastics and biomass appear to be a more sustainable and cleaner alternative.

Finally, the feasibility of this process was investigated by conducting experiments on a 2.5 kW thermal bench scale unit and simulating it using Aspen Plus software. The coherence between the experimental and simulated data was verified, indicating that the process is feasible and can be scaled up for commercial applications.

571p Dopant Modification to Vanadium Phosphorus Oxide Redox Carrier for Chemical Looping Methanol to Formaldehyde Conversion

Sudeshna Gun, Falguni Akulwar, Sonu Kumar, Anuj Joshi, Liang-Shih Fan

Formaldehyde is the key intermediate for numerous high-demand products like resins, polymers, adhesives, and paints, making it one of the most valuable chemicals in the world. Current industrial facilities produce formaldehyde from methanol via the silver catalyst process and the Formox process which involve the passage of methanol and air over a catalyst bed. Despite the widespread adoption of these commercial processes, significant challenges persist. The hazardous methanol-air mixture involved in these processes limits the throughput of the product, and the continuous exposure to reactant gases leads to catalytic deactivation. Hence, with an increase in demand for these formaldehyde-derived chemicals, several studies have been conducted to develop processes for efficient production of formaldehyde. A recent study employed a 2-reactor chemical looping scheme where the molecular oxygen of the redox carrier (vanadium phosphorus oxide) selectively oxidizes methanol to formaldehyde. The reported 85% conversion of methanol and the 45% selectivity towards formaldehyde is significantly less than the commercial processes, impacting its financial viability. Hence, the present study investigates dopant modifications aimed at enhancing carrier performance for methanol conversion to formaldehyde. Metal dopants were shortlisted by comparing their effectiveness in reducing the carriers under methanol over multiple redox cycles. A subsequent analysis was performed to assess the impact of dopant concentration on oxygen carrier performance. Furthermore, this research explores the influence of dopant modification on the oxygen carrier phase using X-ray diffraction analysis. A near 15% enhancement in redox carrier performance has been observed over 10 redox cycles for iron-doped vanadium phosphorus oxide. Further analysis will be carried out to improve the carrier performance and stability over multiple redox cycles. Fixed bed studies will be conducted over the doped oxygen carrier to assess the improvement in methanol conversion and selectivity for formaldehyde. Overall, this study assesses the dopant modification strategy to enhance the performance of the chemical looping scheme for selective methanol oxidation to formaldehyde.

571q Silver Nanoparticle Sensor Array for the Detection of Sars-Cov-2

Benjamin Lam

We report a silver nanoparticle (AgNP) sensor array to detect SARS-CoV-2 viral particles utilizing five peptide receptors to produce a colorimetric response. We show that the interaction of the virus particles with the peptides interferes with the shift in plasmonic absorbance and hinders the formation of snowflake-like fractal assemblies between the AgNPs and the bridging peptides. The limit of detection in water of the sensor array was 500,000 viral copies/mL. We demonstrate the capabilities of the sensor array to distinguish between SARS-CoV-2 (Beta, Delta, and Omicron variants) from the influenza virus at the 99% confidence level through linear discriminant analysis. The sensor array was also able to discriminate three out of five negative exhaled breath condensate samples from five positive SARS-CoV-2 EBC samples.

SARTORIUS

Simplifying Progress

Stability of mRNA and Lipid Nanoparticles Stored in Single-Use Bags

Yuji Takeda[1], Fujun Wang[1], Nina Perier[2], Xin Gu[1], Lucie Delaunay[2], and Samin Akbari[1]
[1] Sartorius Stedim North America Inc., Marlborough, MA [2] Sartorius Stedim FMT S.A.S., Aubagne, France
* Corresponding author: Samin.Akbari@sartorius.com

Summary

RNA therapy holds immense promise in revolutionizing the treatment landscape for various diseases. For more effective delivery of mRNA to cells, mRNA is encapsulated in lipid nanoparticles (LNPs). During the manufacturing process, mRNA and LNPs need to be stored in storage bags in an ultra-low freezer. In this presentation, we studied the stability of mRNA and mRNA-encapsulated LNPs stored in single-use bags. We prepared LNPs using our lab-made microfluidics mixing device, which enables us to prepare LNPs at a rate as fast as 120 mL/min. We report that mRNA-loaded LNPs were stable for at least 6 months at -80°C in Flexboy® bags and that in vitro transfection activity was confirmed using the Incucyte® AI Cell Health Analysis Software Module. We also found that the count rate for dynamic light scattering (DLS) dropped within the first hour of storage, suggesting potential particle adsorption to the storage container. The stability of mRNA and mRNA-encapsulated liposomes (in vivo-jetRNA®+) was also reported.

Flexboy® and Flexsafe® Bags

- Flexboy® bags (5 mL to 50 L) and Flexsafe® 2D bags (20 mL to 50 L) have provided robust solutions for the storage, sampling, and transfer of all biopharmaceutical fluids in all process steps.
- These bags promise quality supply, robustness, closure integrity, biocompatibility, and cell growth performance.

Flexboy® bags

Flexsafe® bags

LNP Stability in Flexboy® Bags

- Confirmed stability of LNP in Flexboy® bags at -80°C at least for 6 months.
- Flexboy is suitable for cryostorage of LNP.
- Observed drop of count rate by ~15% in the first hour, suggesting potential particle adsorption to the storage containers.

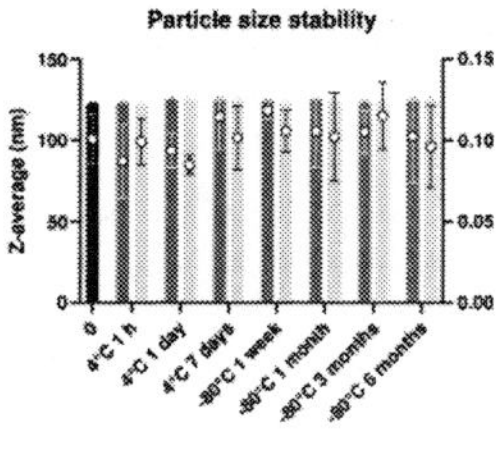

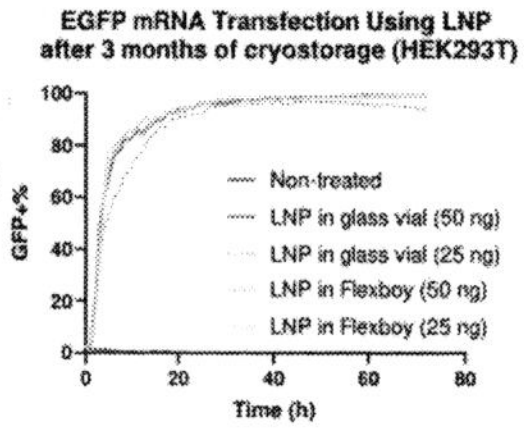

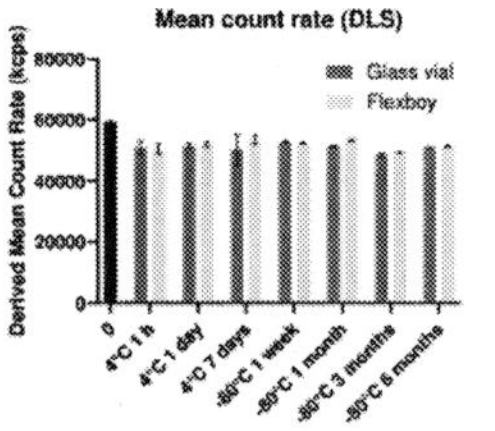

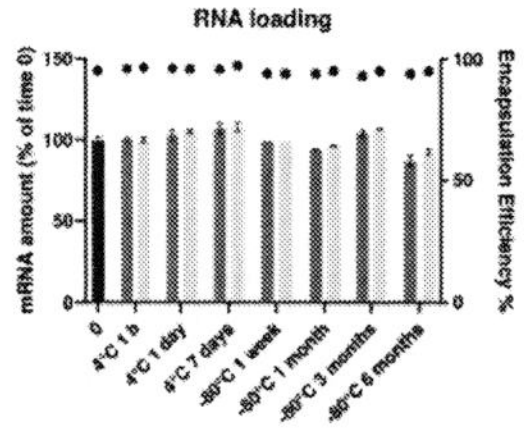

LNP Stability in Flexsafe® Bags

- Confirmed stability of LNP in Flexsafe® bags at 4°C at least for 1 week.

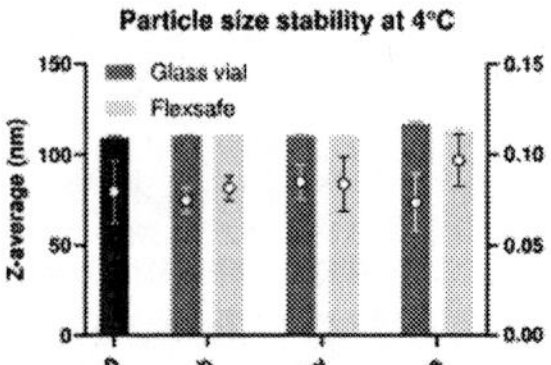

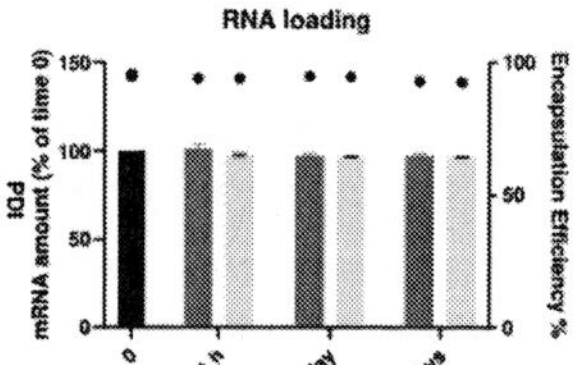

Materials and Methods

- LNP, made of ALC-0315, cholesterol, DSPC, and ALC-0159 lipids, was prepared using a microfluidic device. EGFP mRNA was encapsulated in the LNP. Sample was dialyzed against Tris buffer with 10% sucrose and then sterile filtered. Particle size and count rate were measured using Zetasizer, and mRNA loading was measured using RiboGreen assay.
- Transfection assay: Fluorescent images were taken using IncuCyte® S3 and analyzed using IncuCyte® AI Cell Health Analysis Software Module.

Membrane Microfluidics for Nanoparticle Synthesis

- Membrane chip consisting of micro-channels, track-etched membrane, and pressure-sensitive adhesives.
- Membrane pore size and mixing channel width can be tailored as needed.
- Scalable (up to 120 mL/min) production of LNP with consistent critical quality attributes (CQAs).

Membrane mixer chip

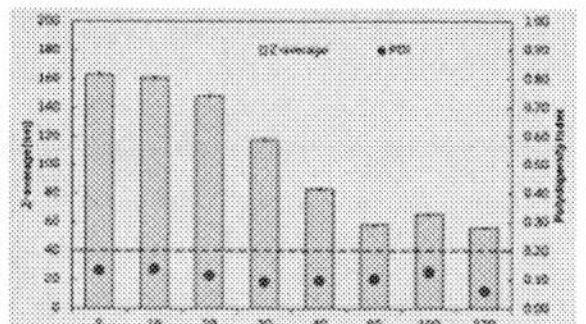

Transfection assay after 3-month storage of LNP at -80°C (HEK293T cell).

	Non-treated	LNP in Flexboy (25 ng RNA)	LNP in Glass Vial (25 ng RNA)
3 h	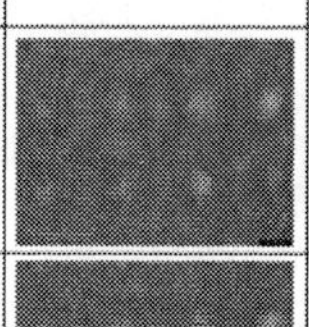	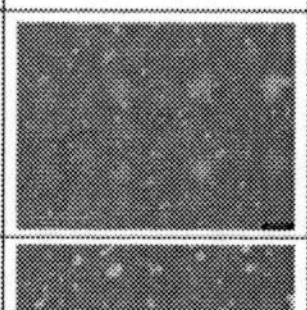	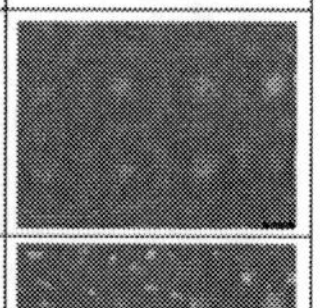
12 h			

Stability of *in vivo*-jetRNA®+ in Storage Bags

- Confirmed stability of mRNA-encapsulated in vivo-jetRNA®+ in Flexboy® and Flexsafe® bags at 4°C at least for 2 weeks.
- Observed drop of count rate by ~15% in the first hour, suggesting potential particle adsorption to the storage containers.

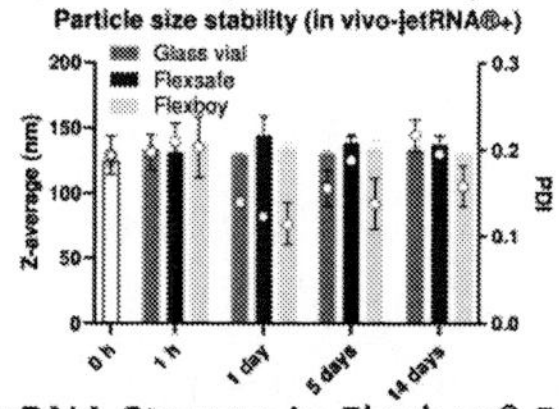

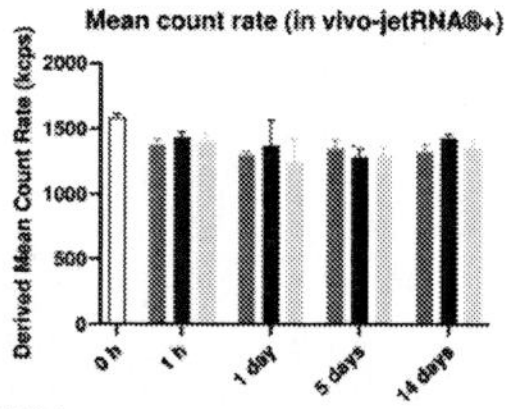

mRNA Storage in Flexboy® Bags

- Confirmed stability of mRNA in Flexboy at -80°C at least for 4 months.
- Flexboy is suitable for cryostorage of mRNA.
- Adsorption of mRNA to the storage containers was not observed.

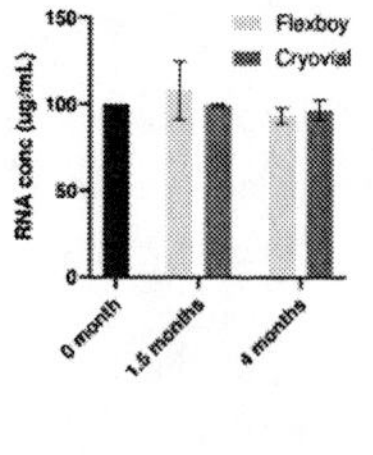

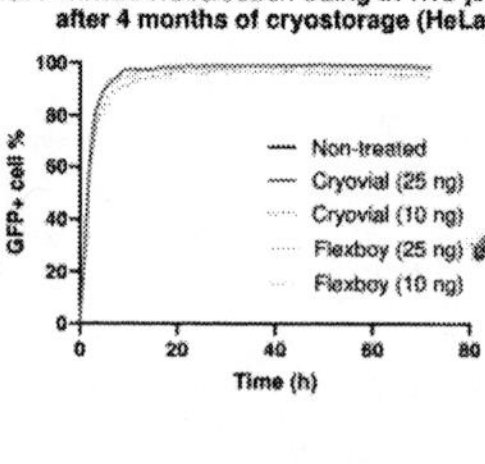

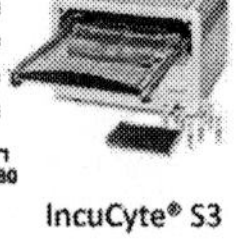

IncuCyte® S3

571s The Effect of Fines and Process Conditions on Particle Abrasion

Alejandro Gallegos Tovar

In flowing granular systems, such as conveying lines, fluidized beds, and hoppers, particles are subjected to different stresses which eventually lead to abrasion, or loss of small surface asperities. Abrasion impacts several key aspects of industrial operation, performance, and design, and often leads to environmental problems and the loss of valuable particle mass. In this study, particle abrasion is examined for three materials (pumice, bauxite, and FCC) in two different flowing granular systems. The first system is a FT4 shear cell in which the effects of increasing concentration of fines, as the particle abrades, and normal force are probed. When a normal force is applied at a constant shear rate, the generation of fines grows exponentially with time. However, if the fines in the shear cell are periodically removed, the abrasion rate is linear, indicating a cushioning effect of the fine particles on abrasion. The second system is a single-pass impact system in which a gas jet laden with particles impacts a flat plate, keeping the total solids loading for each pass constant. Single-pass systems allow for better assessment of the effect of fines, as the fines impact the plate the same number of times as the coarse material. In this system, the effect of fines concentration, gas velocity, particle size, and solids loading are investigated. Particle size distributions and microscopic images of the particles before and after abrasion, and high-speed video of particles in the near wall region, are also obtained. To isolate the effect of fines on particle abrasion, fine particles (under 44μm) are periodically removed after a given number of passes. Particle abrasion, as opposed to particle breakage, occurs below a critical gas velocity, and, without fines removal, the abrasion rate decreases as the number of passes increases.

571t Eco-Friendly Dual-Shell Fungal Chitosan-Silica Microcapsules with Enhanced Mechanical and Barrier Properties for Potential Consumer Applications

Daniele Baiocco, Mohammed Al-Sharabi, Benjamin Lobel, Olivier J. Cayre, Alexander Routh, Zhibing Zhang

Nowadays, heightened self-awareness has swept across the world, resulting in a surging demand for wellbeing, personal care, and cosmeceutical products that positively impact consumers' self-esteem [1]. To this end, a plethora of clean beauty products (CBP) has emerged, with particular emphasis on leave-on/rinse-off formulations enriched with microparticles for skin exfoliation and fragrance delivery properties [2]. Indeed, many CBP have incorporated synthetic microparticles intentionally to maximise the benefits on consumers' skin for a smoother texture, scented touch, and defence against wrinkles and blemishes. Although highly effective, these CBP microparticles, which may fall under the category of microplastics, have posed a severe threat to the ecosystems when rinsed off into wastewater [3]. Following the proactive stance by the European Union (EU) towards banning microplastics from cosmetic products in favour of greener materials, the search for high-performance cosmetic alternatives for skin scrubbing and fragrance delivery is still ongoing.

In this study, we propose a facile method to fabricate fungal chitosan-silica microcapsules with a core of floral scented hexyl salicylate (HS) as a model cosmetic oil. Specifically, emulsified HS droplets were stabilised electrostatically and sterically using sodium dodecyl sulphate (SDS) and SiO_2 nanoparticles (NPs) with their surface modified by ε-polylysine, respectively. Fungal chitosan was used to form an organic coating (primary shell) around the oil droplets at pH 2, which was then reinforced by a secondary inorganic coating of SiO_2 *via* external alkaline (pH 11) mineralisation of sodium silicate. The microcapsules were assayed for their morphological, barrier, and mechanical properties by bright field optical/fluorescence microscopy, scanning electron microscopy, UV-Vis spectrophotometry, and a micromanipulation technique, respectively.

The findings showed that single-shell fungal chitosan microcapsules were spherical with a core-shell configuration and a Sauter diameter of 42.3±0.4 μm, whereas dual-shell microcapsules had a mineralised crown made of SiO_2 (on top of the chitosan shell) leading to a greater Sauter diameter averagely (51.4±0.4 μm). After 15 days in neutral pH water, only 2.5% was released from dual-shell microcapsules, while single-shell microcapsules released ~10% cumulatively. Under compression, dual shell microcapsules yielded a mean nominal rupture stress of 3.0±0.2 MPa, which is statistically much higher than that of single shell ones (1.7±0.1 MPa). This demonstrated that the additional SiO_2 coating enhanced the mechanical properties of the microcapsules significantly, which may be appealing for commercial skincare applications.

The results suggest that green perfume delivery microcapsules with a dual-shell organic-inorganic shell can be effectively fabricated, leading to enhanced mechanical properties and better barrier properties, with potential applications in multiple daily cosmetic products. Future works will focus on assessing the mechanical resistance and fragrance delivery of these microcapsules on soft skin-mimicking substrates. The details of the study shall be presented.

References

[1]. Baiocco, D., J.A. Preece, and Z. Zhang, *Microcapsules with a Fungal Chitosan-gum Arabic-*

maltodextrin Shell to Encapsulate Health-Beneficial Peppermint Oil. Food Hydrocolloids for Health, 2021: p. 100016.

[2]. Anagnosti, L., A. Varvaresou, P. Pavlou, E. Protopapa, and V. Carayanni, *Worldwide actions against plastic pollution from microbeads and microplastics in cosmetics focusing on European policies. Has the issue been handled effectively?* Marine Pollution Bulletin, 2021. **162**: p. 111883.

[3]. Tagg, A.S. and M. Labrenz, *Closing Microplastic Pathways Before They Open: A Model Approach.* Environmental Science & Technology, 2018. **52**(6): p. 3340-3341.

38a Systematic Study of Caking in Sintering and Soft Surface Systems – the Role of Particle Size, Temperature and Creep Effects

Kerry Johanson

One problem of working with bulk materials is that they form very strong agglomerates, lumps, or even large solid masses in process equipment. We call this problem caking. But, caking is also a mechanistic problem and has multiple causes that result in bulk material gaining strength after storage. Understanding caking requires an understanding of the kinetics that happen between particle-particle surfaces. Often, we lump all caking into similar kinetics explained by the rate of formation of some binding material in a confined space as happens in crystallization systems. However, some caking mechanisms occur when the bulk surface of the material deforms and develops welding events between adjacent particles through warm solid creep events. The kinetics of this type of caking are driven by the formation of new surfaces through particle deformation. This suggests that the creep behavior of a bulk material should have similar kinetics to the kinetics that describe the gain of strength of the bulk material. Often these types of caking events occur as the material is subjected to a single temperature excursion for some critical time. Associated with these systems is a significant dependence on the temperature where the bulk unconfined yield strength becomes very large while the material is stored at some critical temperature. This cohesive behavior increases exponentially as the critical temperature is approached.

From a continuum point of view, unconfined yield strength is defined as the major principal stress that causes a bulk material to yield in shear while in an unconfined state after it has experienced a consolidation stress causing it to compact. From a particle viewpoint, during a shear event adjacent particles are in the process of two distinct actions. Some adjacent particles slide past each other, causing frictional resistance forces. Some adjacent particles adhere to each other, and during a shear event, these particles pull apart and rupture the adhesion connection. So, the bulk unconfined yield strength (a fundamental continuum property) is a function of the collection of all the friction forces as well as all the adhesion forces between adjacent particles.

The unique feature of this mechanism is that the adhesion of the surface depends not only on the temperature of the surface, but also on the temperature of the entire mass of the particles. The surface can become sticky with increasing temperature but, over time, the area involved in adhesion also changes because the shape of the adjacent particles change, resulting in larger more intimate connections between particles through local deformation of the particle surfaces. In some cases, this deformation leads to cases where the intruding particles deform in such a way that a divot is formed in the adjacent particle matching the shape and size of part of the intruding particle. Thus, these particles interlock just like puzzle pieces interlock. When shear begins the divot resists the shear event due to the fact that, in order to shear, the adjacent particles must first move up out of the divot and then slide past each other. This increases the bulk unconfined yield strength of the material. This work is a systematic study of the events that influence the interaction of surfaces that are prone to local deformation and caking due to warm creep effects. The work looks at the relationship between the caking and the creep effects. It also looks at the relationship between the change of surface area contacts over time and the resulting caking effects. Finally, it looks at the role of the induced local deformations in preventing shear between adjacent particles during failure in an unconfined state.

38b Quantifying Cohesive, Frictional and Interlocking Effects for Universal Powder Flow Characterisation

Amalia Thomas

The infinite combinations of powder processing and handling conditions make it difficult for a single powder flow characterisation instrument or testing methodology to accurately simulate how a material will perform in any flow pattern in any combination of energy, shear-rate and stress regimes. Traditional and newer rheological testing methodologies can facilitate a quantitative and sensitive differentiation of powder samples, but the root morphological cause of behavioural differences between the samples is often unknown, and the various testing methodologies can lead to conflicting results.

We explore a new framework of interpretation of a minimal number of rheological tests for powder flow characterisation based on the quantitative description of the mechanisms that oppose the motion of particles past each other: cohesion, friction and mechanical interlocking. The degree to which each mechanism contributes to 'flowability' scales differently with the normal force between particles and with the shear rate induced, and the different mechanisms lead to different flow-related issues: while a highly cohesive powder will tend to form clumps and agglomerates and loosely packed structures, a highly frictional and interlocking powder will form strong force chains than can arch and jam at openings. The proposed methodology offers numerical values that inform directly on the effects of morphological and environmental conditions, and that relate directly to powder performance in a wide variety of powder processing and handling scenarios.

We propose quantifying the relevance of cohesive forces on powder flow through the ratio of inter-particle cohesive forces to the average particle weight, i.e. the granular Bond number, Bo. This number then defines the degree to which cohesion prevents particles from fitting into gaps and flowing past each other due to gravity, leading to a poor packing efficiency and a likelihood of clumping. We propose the most direct method for measuring Bo can be carried out through a recently developed Fluidization test, where the fluidization behaviour and uniformity of a powder depends on the balance of air drag, particle weight and cohesive forces – by measuring the excess pressure drop across the bed necessary to fluidise a powder, relative to the pressure drop once it is fluidised, we calculate how much larger the net interparticle cohesive force is than the average particle weight, and so Bo. The difficulty of layers of powder to shear over each other is quantified through the coefficient of friction obtained by shearing layers of powder under different applied loads in a Shear Cell test, while mechanical interlocking effects due to particle shape irregularity are calculated based on the powder packing efficiency and on how this changes under compressing stresses.

By quantifying the intrinsic strength of cohesive bonds, the effective coefficient of friction and the packing fraction under different consolidating loads, the degree to which each mechanism contributes to 'flowability' can be predicted in terms of process parameters. This information can complement other rheological testing by rationalising differences in sensitivity and 'flowability'-ranking obtained through other methodologies, it can support a more accurate understanding and prediction of how the powder will behave in different conditions, and it can inform on the best course of action for solving flow-related issues. The metrics used are all dimensionless, allowing for the first time, a universal direct comparison of flowability across completely different materials, and a common language for powder flow characterisation across materials, processes and industries.

38c Insights into Vibrational Rheology for Determining the Stickiness of Food Powders

Maria Cares

Food powders are complex systems with diverse ingredient formulations, entailing numerous physicochemical interactions among particles, containers, and the surrounding environment. Industrial processes that handle such powders often encounter challenges related to their rheological properties. These properties are susceptible to environmental factors like relative humidity and temperature, which can lead to stickiness and caking phenomena, compromising storage and processing quality. Understanding and predicting stickiness in food powders is crucial for various stages of production and consumption. However, the field currently lacks standardized measurement techniques and predictability. Despite advancements in assessing powder stickiness over the past decades, persistent challenges remain in controlling environmental conditions during testing, accounting for surface phenomena sensitivity, and considering the kinetics of the process.

This study introduces a novel method for assessing powder stickiness by examining its kinetics, particularly focusing on the onset of stickiness (t_s) in amorphous food powders through rheology. The method utilizes a rheometer equipped with a shaker and a cell for in-situ moisture control, facilitating fast and automated assessment of powder flow behavior. By using rheology to identify the point at which the powder's elasticity and viscosity significantly increase, our method provides insights into the onset of stickiness.

In contrast to traditional devices requiring temperature adjustments and lengthy equilibration, our powder cell, equipped with humidity control, enables in-situ examination of water uptake at ambient temperature. Method validation extends across various powders, including glass beads, maltodextrins, and soluble coffee. Vibration within the system continuously reorganizes particles, improving measurement accuracy, ensuring homogeneity, and heightening sensitivity to surface properties.

Results demonstrate that powder stickiness is a complex kinetic phenomenon involving various surface and volume mechanisms, such as moisture diffusivity, plasticization, liquid bridge formation, and cohesion, dependent on both the powder's chemical (functionality, surface chemistry) and physical (particle size distribution, rugosity, and morphology) nature.

Our approach offers valuable insights for understanding and quantifying stickiness kinetics in food powders, potentially enhancing food processing and quality control. It operates efficiently, minimizing powder attrition, making it suitable for ambient conditions, and bridging the gap between the behavior of powder and environmental conditions in the food industry.

38d Air Permeability of Powders at Process Conditions

Marco Lupo, Geoffroy Lumay, Aurelien Neveu, Filip Francqui

The powders are widely used in many industrial processes, e.g. pharmaceutical, food, battery, additive manufacturing. Many are the properties of the powders that influence the efficiency of the processes and the quality of the final product. Among these properties is the permeability, which can be defined as the capacity of a powder to transmit a fluid, usually air, through its bulk. For example, permeability affects: 1) the filling weight in dosator nozzle machines used for capsule filling (Faulhammer et al., 2014), 2) the effective dispersion and release of the inhalation mixture of dry powder inhalers (Shalash et al., 2018), 3) the capacity of the powder to get out from the silo during powder silo discharges (Hsiau et al., 2012).

Depending on the process conditions, the powder reaches specific packing fractions. The packing fraction, defined as the ratio between the bulk and the true density, is a measure of the powder densification. The permeability is strongly dependent on the packing fraction, as the larger the powder densification, the lower the permeability. On the contrary, the looser the powder, the easier the passage of the air through the particles and hence the larger the permeability. Therefore, it is important to measure the permeability at the same packing fractions induced to the powder by the process to properly correlate the permeability to the quality of the final product. In this work, a new measurement method is proposed. It combines the possibility of selecting the packing fraction in a wide range of densification conditions and the possibility of measuring the permeability in correspondence with the chosen packing fraction values. The instrument adopted is the GranuPack (Granutools, Belgium), a high resolution tapped density analyzer, by which it is possible to control the powder densification by applying the desired number of taps to the powder, contained in the permeability cell. At the bottom, the cell presents a porous frit to prevent the powder from entering the air supply pipe but allow the passage of the air for the permeability measurement. By alternating the application of taps and the permeability measurements, it is possible to have access to the evolution of the permeability with the packing fraction. Different lactose and abrasive powders, characterized by different shapes and sizes, have been tested and the evolution of permeability with the packing fraction has been obtained. Two parameters, that quantify the effect of densification on permeability, are proposed: the permeability ratio and the rate of variation of permeability. These parameters are directly related to the powder cohesiveness and hence can be used to complement the classical flowability indexes.

The new insights gathered from this approach give a better understanding of the material properties and allow the development of improved formulations for better processability.

References

Faulhammer, E., Llusa, M., Radeke, C., Scheibelhofer, O., Lawrence, S., Biserni, S., Calzolari, V., & Khinast, J. G. (2014). The effects of material attributes on capsule fill weight and weight variability in dosator nozzle machines. *International Journal of Pharmaceutics*, *471*(1–2), 332–338. https://www.sciencedirect.com/science/article/pii/S0378517314004116

Hsiau, S., Liao, C., & Lee, J. H. (2012). The discharge of fine silica sand in a silo under different ambient air pressures. *Physics of Fluids*, *24*(4). https://pubs.aip.org/aip/pof/article/24/4/043301/257578

Shalash, A. O., Khalafallah, N. M., Molokhia, A. M., & Elsayed, M. M. A. (2018). The Relationship

Between the Permeability and the Performance of Carrier-Based Dry Powder Inhalation Mixtures: New Insights and Practical Guidance. *AAPS PharmSciTech*, *19*(2), 912–922. https://doi.org/10.1208/S12249-017-0898-7

38e Investigating the Impact of AlO_x Atomic Layer Deposition on the Properties and Flow Behaviour of ZSM-5 Zeolites Toward Improved Powder Processing

Angeliki Chalasti, Canan Gücüyener, Ian Harkness, Jerry Heng

Zeolites are a class of microporous crystalline materials with regularly arranged pore structures and tunable chemical functionalities. These exceptional properties make them extremely valuable for various industrial applications, including catalysis, separation, and adsorption [1][2]. Throughout their production process, the properties of zeolite undergo significant transformations influenced by various particle handling techniques, including grinding, calcination, drying, storage, discharge, and transport, alongside specific processing parameters [3]. Understanding the flow behaviour of zeolites becomes imperative for optimizing these procedures, thus enhancing the quality of the final product, and maximizing process efficiency. While the influence of particle properties like particle size and shape on powder flow is well-studied, identifying the interplay between different factors governing powder behaviour remains elusive. In this aspect, the surface attributes of particulate solids, such as chemical composition and surface roughness, play a profound role in powder flow behaviour [4-6]. Nevertheless, there is a lack of comprehensive understanding regarding how surface modification and the resulting alteration of material properties can influence interparticle interactions within zeolite powders, with limited prior research ever dedicated to this aspect.

Over the past few years, Atomic Layer Deposition (ALD) has emerged as a powerful tool for modifying the surface characteristics of catalytic substances like zeolites [7]. ALD is a self-limiting method for atomic-level deposition of nanoparticles and film growth. Depending on the desired application ALD is often applied to regulate zeolite properties such as their surface acidity, defect concentration, and hydrophilicity. In this work, we aim to investigate how the ALD process of AlO_x on ZSM-5 zeolites affects the particle properties and its consequent effect on powder flow behaviour to leverage the understanding of the multifaceted factors governing this process.

AlO_x was deposited on the parent ZSM-5 to create four samples of increased layering (1, 2, 5 and 10 layers) and the materials were examined and referenced to the bare zeolite using a variety of advanced characterisation techniques. Inverse gas chromatography (IGC) was employed to determine the surface energetics of the samples through the injection of a series of non-polar (pentane, hexane, heptane) and polar (dichloromethane and ethyl acetate) probes [8]. The moisture sorption characteristics of the samples were examined at relative humidities ranging up to 90%, using dynamic vapour sorption (DVS) [9]. A surface elemental analysis was performed using X-ray photoelectron spectroscopy (XPS), and the results were compared with the bulk composition of the materials using X-ray fluorescence (XRF). N_2 adsorption for BET surface area and pore volume determination, true and bulk density measurements, particle size analysis, and scanning electron microscopy (SEM) were also utilised to comprehend the alterations in physicochemical characteristics and surface chemistry induced by the ALD process. Moreover, the flowability of the five powders was examined using an Anton Paar shear cell for consolidation stresses ranging from 1 kPa to 5kPa. The experiment obtained stress data for normal and shear stress at different incipient failure points and performed shear tests at three different applied normal stresses - 30%, 50%, and 70% of the pre-shear normal stress - to determine the yield locus of each powder sample. This enabled the calculation of the flow function coefficient (ff_c) at different applied pressures. A statistical analysis was carried out using JMP software to examine the correlations between the physicochemical and flow properties of the samples.

It was found that the surface chemistry modification resulting from the ALD significantly affected the bulk and surface properties of the samples, leading to an improvement in powder flow with the increase in AlO_x layering (Figure 1). The ALD treatment caused a decrease in the total surface area, the micropore area and the pore volume of the zeolites (Figure 2). The parent sample possessed a surface area of 407 m^2/g, which was reduced by 13% with the 10th layer coating. Similarly, micropore area and pore volume decreased by 11% and 17%, respectively. This change was also reflected in the reduction of the moisture uptake of the samples with the increase of AlO_x layers. Furthermore, the ALD modifications led to a substantial reduction in both surface silica to alumina ratio (SAR) and bulk SAR, with a notable decrease of 97% and 56%, respectively, following the addition of the 10th layer (Figure 3). While primary particle size and particle morphology seemed to remain unaffected after the ALD, the introduction of AlO_x appeared to weaken interparticle cohesion, as evidenced by variations in the particle size distribution among the samples. These findings prompt ongoing investigation through surface energy studies to provide deeper insights into the observed effects.

It was found that there is a strong positive relationship (r=0.88) between the number of AlO_x layers and the flow function coefficient (ff_c). Strong correlations have also been established between various zeolite surface properties, and their powder and bulk characteristics (Figure 4). For instance, the surface area negatively correlated with the ff_c at 1kPa of consolidation stress (r=-0.88). The decrease in surface area decreased the contact area between particles, hence reducing interparticle friction and facilitating flow. Similarly, the pore volume and moisture uptake at 90% RH developed a negative trend with the ff_c (r=-0.83 and r=-0.94). These results suggest that decreasing the pore volume of zeolites and, as a result, their water sorption capacity potentially lessened the available space for water molecules to accumulate, minimizing the capillary forces among particles and improving the flow. Further experiments are currently being conducted to confirm the observed trends.

In summary, this research underscores the role of the ALD modification on the zeolite's particle properties and in determining powder flowability, which can play a pivotal role in optimizing industrial processes and improving product quality. Increasing the ALD coating resulted in a significant alteration of the physicochemical attributes of zeolites, thereby enhancing their powder flowability.

References

[1] E. Pérez-Botella, S. Valencia, F. Rey, Zeolites in adsorption processes: State of the art and future prospects, Chemical Reviews 122 (2022) 17647–17695.

[2] Y. Li, L. Li, J.Yu, Applications of zeolites in sustainable chemistry, Chem 3 (2017) 928–949.

[3] S. Zinatlou Ajabshir, C. Gucuyener, V. Vivacqua, D. Gobby, H. Stitt, D. Barletta, M. Poletto, Flow behaviour of zeolite powders at high process temperatures, Powder Technology, 409 (2022) 117818.

[4] C.G. Jange, R.P.K. Ambrose, Effect of surface compositional difference on powder flow properties, Powder Technology, 344 (2019) 363–372.

[5] U. V. Shah, V. Karde, C. Ghoroi, J. Y.Y. Heng, Influence of particle properties on powder bulk behaviour and processability, International Journal of Pharmaceutics, 518(1–2) (2017) 138–154.

[6] V. Karde, S. Panda, C. Ghoroi, Surface modification to improve powder bulk behavior under humid conditions, Powder Technology, 278 (2015) 181–188.

[7] T. Gong, L. Qin, J. Lu, H. Feng, ZnO modified ZSM-5 and Y zeolites fabricated by atomic layer deposition for propane conversion, Physical Chemistry Chemical Physics, 18(1) (2016) 601–614.

[8] F. Thielmann, D.J. Burnett, J.Y.Y Heng, Determination of the surface energy distributions of different processed lactose, Drug Development and Industrial Pharmacy, 33(11) (2007) 1240–1253.

[9] J.Y.Y. Heng, D.R. Williams, Vapour Sorption and surface analysis, Solid State Characterization of Pharmaceuticals, (2011) 245–285.

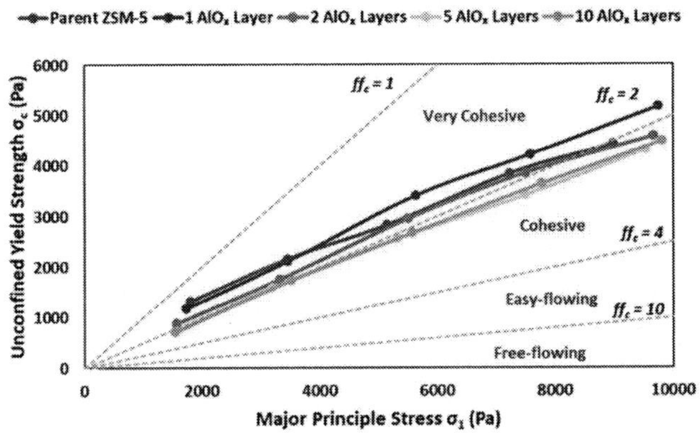

Figure 1: Flow function coefficients of parent and atomic-layer deposition ZSM-5 using the Anton Paar Shear Cell.

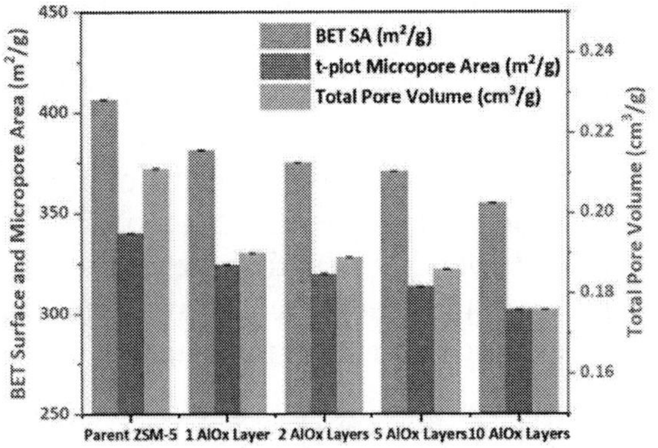

Figure 2: BET surface area, micropore area and total pore volume of parent and atomic-layer deposition samples.

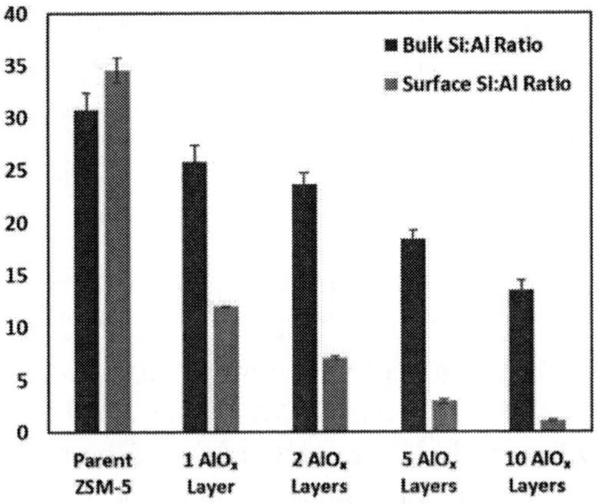

Figure 3: Comparison of bulk and surface Si:Al ratio (SAR) with increasing AlO$_x$ layers for parent and atomic-layer deposition samples.

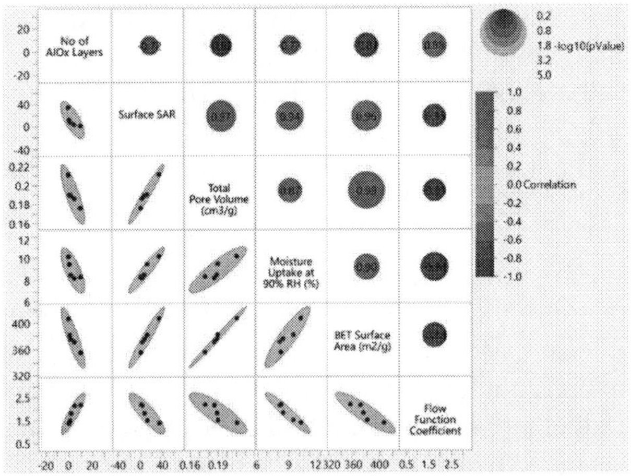

Figure 4: Correlations scatterplot matrix of surface and bulk properties in relation to no of AlO$_x$ layers and ff_c showing the correlation coefficients (r values). The size of the circles indicates the strength of the correlation, with positive r-values in red and negative in blue.

38f Prediction of the Spreadability of Metal Powders: The Last Developements

Aurelien Neveu, Filip Francqui, Geoffroy Lumay

The spreadability of powders, i.e. their ability to produce smooth and homogeneous layers, is an essential property to guarantee the good quality of the parts produced in powder-bed based AM processes. However, evaluating the spreadability of new materials during production is usually not feasible due to the cost associated with the minimum batch volume to fill in the machine and the time required to empty and clean the machine between each test.

The Cohesive Index metric (GranuDrum, Granutools, Belgium) has been demonstrated to be well correlated with the homogeneity of the deposited layers in an SLM printer (ISO/ASTM TR 52952:2023). This indicates the strong relationship between powder cohesiveness and spreadability. These results opened up new perspectives to predict the spreadability of materials a the formulation stage, without having to produce large batches. Also, suppressing the need to fill the machine to evaluate the spreadability is nice to have to reduce the material cost and machine time. Recently, the study has been extended to broaden the range of types of material and particle sizes and confirmed the good correlation between the Cohesive Index and layer spatial homogeneity (Francqui et *al.*, ICAM2023).

Investigating the spreadability directly in the SLM machine has the advantage of being close to the real configuration in production. Indeed, the actual spreadability is influenced by the powder property but also by the recoater geometry (blade/roller), the wiper shape and material (polymer, metal), and the bed roughness especially for the first layers. However, the instrumentation is restricted by the build chamber volume and the moving parts. In the previous studies, the built-in video camera system was used to acquire pictures of the powder bed. The pictures were then analyzed with a custom image processing algorithm to get a measure of the layer spatial homogeneity, the interface fluctuation metric.

To tackle some of the limitations of the in-situ evaluation, a test bench has been developed and instrumented (LEM3, Metz, France). The mechanical parts (bed, recoater) of the test bench are taken from the real machine (SLM280, SLM Solutions, Germany). Therefore, even if the evaluation is not performed inside the machine, the experimental conditions are still very close. Powder layers are sequentially deposited, and the layer quality is evaluated between each layer. The device has been instrumented with a confocal microscope used to precisely measure profiles of the height variation across the layer. The layer local topology is thus directly accessible. In addition, powder samples are taken from different locations of the spred and the particle size distribution is analyzed.

This measure is complemented with optical images taken by a camera placed orthogonally to the powder bed. Moreover, the lighting system is designed so that the light orientation can be varied relative to the recoater direction. Indeed, we show that depending on the light direction, the observation of defects attributed to the powder cohesiveness can be amplified compared to the ones attributed to the interaction with the recoater.

38g Influence of Particle Properties and Moisture Content on the Flowability of Microcrystalline Cellulose Powder

Jordan Monroe, Heather Emady

Understanding particulate flow is crucial to the design and development of solids handling in industries from foods to pharmaceuticals. The relationship between particle properties (micro-scale), bulk particle flow (meso-scale), and for a broad range of powders is not well established due to the complexity of solids systems.

This work experimentally investigates the effects of particle size and moisture content on the flowability of micro-crystalline cellulose (MCC). The raw MCC was sieved into varying size fractions based on the mesh size ranges of 0-25 µm, 25-45 µm, 45-63 µm, 63-90 µm, and 0-90 µm. Particles in each size fraction were characterized by optical microscopy via a Malvern Morophologi G3SE Particle Analyzer for their and shape descriptors. Using a Freeman FT4 Powder Rheometer, bulk powder properties to characterize flowability were found, including basic flowable energy, compressibility, flow function coefficient, etc. Moisture content was varied via the addition of deionized water in amounts from 0% to 50% by total weight. Particle size, shape, and moisture content were correlated with the FT4 flowability results. This is the first study, to our knowledge, to quantify the effects of particle shape, size, and moisture content on the flowability of powders.

280a Conformal Protective Coating of Cathode Materials By Oxidative Chemical Vapor Deposition for Enhancing Lithium-Ion Batteries.

Mehran Seifollahi, Kenneth Lau

Relevance and Problem

Lithium-ion batteries (LIBs) are relatively energy dense, which make them suitable in energy-demanding applications such as electric vehicles and grid storage. To further increase energy density and cycle life, highly active cathode materials are being pursued, such as layered lithium transition metal oxides for the battery cathode. Despite their relatively high capacity, layered lithium transition metal oxides (e.g. nickel-cobalt-manganese or NCM) undergo crystal and interfacial degradation when subject to aggressive electrochemical and thermal forces, leading to a rapid drop in performance and poor cycle life. These problems are more severe under more stressful battery operating conditions, such as high voltage and high temperature cycling, which can lead to parasitic side reactions on the cathode surface. The failure mechanisms include interfacial degradation, electronic and ionic isolation, undesired cathode-electrolyte reactions, and transition metal dissolution that lead to impedance increase and capacity loss. Thus, the interfacial challenges of cathode materials need to be addressed to achieve longer cycle life. A proposed solution to overcome these challenges is to use a thin film deposition process to protect the electrode surface, which is a direct means to increase the longevity of cathode materials [1,2].

In this work, we illustrate the use of oxidative chemical vapor deposition (oCVD) as a thin film polymer deposition technique that allows a conformal protective barrier of a conducting polymer to grow on the surface of the LIB NCM cathode particles. The oCVD process, as a solvent-free technique, overcomes many obstacles posed by liquid-based processing methods, such as non-conformal and non-uniform coatings due to poor wettability, and damage to substrates from liquid solvents. oCVD is a versatile method that can synthesize electrically conductive poly(3,4-ethylenedioxythiophene) (PEDOT) in a solventless environment by enabling the oxidative polymerization of 3,4-ethylenedioxythiophene (EDOT) monomer using vanadium oxytrichloride ($VOCl_3$) oxidant in the vapor phase. The oCVD process enables ultra-conformal and highly uniform PEDOT coating on topologically complex structures. This approach makes oCVD a promising candidate for coating PEDOT as a protective layer on NCM cathode materials to enhance the structural stability, capacity retention, and cycle life of LIBs.

Method and Results

This work studies the encapsulation of NCM111 cathode materials with PEDOT coatings using a novel rotary oCVD reactor system that allows the coating of dry cathode particles. The goal is to determine the effect of PEDOT on enhancing LIB performance. The EDOT monomer and $VOCl_3$ oxidant are metered into the oCVD reactor at a flow rate of 2 and 1 sccm (standard cubic centimeters per minute), respectively, to achieve a 2:1 monomer to oxidant flow ratio. The reagents were fed from source jars that were heated to achieve sufficient vapor pressure. The reaction flask is charged with dry NCM cathode particles (1–2 g; 1–10 μm particle diameter) and is continuously rotated at 140 rpm to achieve constant particle agitation and surface exposure to the reagents. The flask temperature is maintained by a temperature-controlled oil bath that was varied from 90 to 130 °C. The deposition time was varied from 60 to 100 minutes to achieve different film thicknesses. The pressure was maintained by a capacitance manometer and butterfly valve, and was kept constant at 300 mTorr for all the trials.

The electrical conductivity of the PEDOT-coated cathode material on an aluminum current collector was measured with a four-point probe. The film conductivity was found to strongly correlate with temperature and deposition time. At a deposition temperature and time of 90 °C and 80 min, respectively, the electrical conductivity reached a maximum of 600 S/cm. Further increasing the deposition time results in a decrease in conductivity, which is attributed to an increase in the density of film defects, impurities such as unreacted reagents, and polymer chain disorder. With the deposition time fixed at 80 min, the effect of the deposition temperature on the electrical conductivity was studied. By increasing the temperature to 130 °C, the conductivity increased dramatically to 5000 S/cm. This is attributed to both a slower film deposition rate and more favorable polymer crystalline orientation. The higher flask temperature lowers the rate of monomer adsorption, which decreases the deposition rate and hence produces a more uniform coating with fewer impurities and film defects. The resulting thinner film also minimizes the increase in impedance caused by the increased diffusion distance of lithium ions across the PEDOT film. In addition, at higher temperatures, the crystalline orientation of the PEDOT film is known to transition from an edge-on orientation to a face-on orientation of the aromatic rings, which enhances the conducting pathway and hence electrical conductivity.

After optimizing the PEDOT deposition conditions to 80 min and 130 °C, the coated NCM particles were assembled into LIB half cells with a Li metal reference anode and liquid electrolyte in an argon glove box. Uncoated NCM particles were also assembled into coin cells as a reference comparison. The cells were formation-cycled at C/10 (20 mA) at room temperature for 5 cycles. The PEDOT coated cells achieve an initial capacity of 220-240 mAh, while the uncoated cells achieve an initial capacity of 180-190 mAh. The higher initial capacity of the PEDOT coated cells is attributed to both the protective layer against electrolytic attack and the PEDOT film acting as active material that can store Li ions. After formation cycling, the coated cells have a starting capacity of 210 mAh, and the uncoated cells have a starting capacity of 160 mAh. The cells were subsequently cycled at 1C (200 mA) between 3 and 4.6 V at room temperature. The coated cells achieve 91% capacity retention after 100 cycles, while the uncoated cells have only 54% capacity retention. Furthermore, additional coin cells were assembled to study the effect of the PEDOT thickness on cycling performance. The thinner PEDOT films provide insufficient protection from the electrolyte, and the thicker films significantly increase the Li ion transport distance, both resulting in lower performance compared to the optimized conditions.

Implications and Future Work

This work shows that oCVD PEDOT can be used to fabricate conductive and protective coatings on NCM cathode particles to protect them from degradation in lithium-ion batteries. This technique can be expanded to other cathode materials that are especially vulnerable to electrolytic attack and suffer from low conductivity. PEDOT coated cathodes have been found to enhance both the capacity retention and cycle life of lithium ion batteries. More broadly, the fabrication of coated cathodes makes lithium ion batteries more practical for high storage and high power applications such as electric vehicles and grid storage, where high-capacity retention and cycle life are simultaneously sought after. Additional work is currently underway to not only coat NCM powders but also coat the assembled cathode structure to encapsulate not only the active cathode particles but also the electrically insulating binder, which should lower series resistance, improve mechanical stability and further enhance cycle life. Furthermore, the coated cells will be tested at more stressful conditions, such as elevated temperatures (>40 °C) and higher currents (>2C), to determine whether the PEDOT coating allows the cells to endure harsher charge-discharge conditions. By optimizing the oCVD processing conditions, batteries with higher capacity and longer cycle life can be manufactured to meet increasing energy storage demands.

References

1.Im, Sung Gap, and Karen K. Gleason. 'Systematic control of the electrical conductivity of poly (3, 4-ethylenedioxythiophene) via oxidative chemical vapor deposition.' Macromolecules 40.18 (2007): 6552-6556.

2.Xu, Gui-Liang, et al. 'Building ultraconformal protective layers on both secondary and primary particles of layered lithium transition metal oxide cathodes.' *Nature Energy* 4.6 (2019): 484-494.

280b Lithium-Sulfur Battery Performance Enhancement through the Integration of Porous Liquid-Based Electrolytes: Insights into Improved Ion Mobility and Ionic Conductivity

Mehran Arzani, Kartikey Sharma, Anika Harishachandra Shinde, Hamidreza Mahdavi, Vikas Berry

Among the alternatives to lithium-ion batteries, the Lithium-sulfur (Li-S) battery emerges as a promising candidate for next-generation energy storage systems, owing to its potential for high energy density, cost-effectiveness, and environmental sustainability. However, several notable challenges hinder its widespread adoption. These include rapid capacity decay, limited cycle life, and the low conductivity of the electrolyte. This study aims to improve the performance and lithium transport conductivity of electrolytes in Li-S batteries using porous liquid-based electrolytes (PLEs) as a potential candidate. The physicochemical properties, solvation mechanism, transport characteristics, and electrical conductivity of lithium ions through PLEs were studied. It was found that PLEs can effectively enhance battery performance by the presence of PLEs' empty, and permanent porosities. This enhancement can be attributed to the capability of coordinating and transporting Li^+ ions with high mobility, and high solvation of Li^+ ions by PLE, which facilitated the dissociation of Li^+-$TFSI^-$ ion pairs and the formation of complexes with Li^+ ions with PLE. The results of this study indicate that the utilization of PLE can present a viable option for enhancing the physicochemical characteristics of electrolytes, thereby contributing to the advancement of Li-S batteries.

280c Dry Process for Battery Electrodes: The Role of Powders Mixture Properties and Mixing Conditions on Flowability and Electrode Performance Using Design of Experiments

Alex Lemarinel, Graciela Cares, Thomas Philippe, Clément Paul, Lauréline Marchal, Philippe Marchal, Veronique Falk

In a rapidly evolving global landscape, lithium-ion batteries (LIB) are increasingly utilized in transportation, energy storage, and electronics underscoring the imperative to enhance battery characteristics. While the near-future LIB production capacity is expected to continue its growth during the next years(1), numerous challenges still need to be faced, for instance, manufacturing requires a better supply chain, including sustainable sourcing of minerals and the use of recycled cells materials. Moreover, the current manufacturing process suffers from several limitations such as high energy consumption and limited speed, both induced by the drying step, which consists in solvent evaporation during heating. LIB dry processes not only avoid the use of toxic solvents, but also aims for breakthroughs in lithium-ion and post lithium-ion technologies contributing to a more environmentally friendly approach, reducing energy consumption and costs while offering a dual environmental advantage. However, the dry process needs the development of specific unit operations such as powder formulation, mixing, electrospray deposition and calendering. Among them, electrospray deposition stands out as one promising technique for dry electrode manufacturing, and it is strongly dependent on the formulation and mixing parameters.

Ensuring large-scale viability necessitates optimizing the dry process from the standpoint of each unit operation. This is crucial for crafting high-performance electrodes, guaranteeing enhanced cyclability and greater capacity during high-rate discharge. The statistical design of Experiments (DoE) is a powerful tool for optimizing processes. This study investigates the impact of formulation and mixing operations on the flowability of the powder mixture and the electrochemical performance of the electrode. The evaluation criteria, DoE responses, include the surface coverage of active particles with the "polymeric binder", minimal fluidization speed, and electrode porosity, all of which seem crucial for achieving high-performance electrodes following electrospray deposition. The DoE factors examined include mixing parameters such as time, rotation speed, and mixing order, along with the investigation of the formulation (active material/PVDF/conductive additives ratios). This methodology serves as an initial approach to analyze which factors are the most influent and to evaluate their impact for the next steps of the process.

[1] « The Battery Report 2023. », Volta Foundation, Jan. 2024. https://www.volta.foundation/annual-battery-report. Accessed 08 Apr. 2024.

280d Powder Characterization for Electrode Production Improvement

Filip Francqui, Salvatore Pillitteri, Geoffroy Lumay, Aurelien Neveu

Electrode production is a critical step, determining the battery properties such as the power density, specific capacity, and longevity. Special care is consequently paid to electrode manufacturing. At the start of production, raw materials, made of active material, conductive additive, and binder, are generally dry-mixed in powder form and then added to a solvent to make a slurry. This slurry is spread on a current collector to create an electrode. In addition, new dry processes are more and more preferred for ecological and financial reasons, directly handling raw materials in dry powder form. Therefore, powder characterization becomes more and more important for electrode manufacturing improvement.

A critical required property for batteries is energy density. The larger the specific and areal capacity, the more energy per unit of weight and volume the battery can store. Batteries based on sulfur as an active material for cathode are more and more considered. However, while sulfur has a high theoretical specific capacity, its poor conductivity makes it unable to exchange electrons during the electric charge. In addition, the poor flowability of sulfur makes it challenging to handle during electrode manufacturing processes. For these reasons, carbon black is generally dry mixed with sulfur powder to improve conductivity and flowability.

Composite powders made of sulfur and carbon black are also envisaged for improving the conductivity and the energy density of the blend. Nevertheless, composite sulfur/carbon black powders give different behaviour compared to blends prepared by dry-mixing. In this work, we analyzed blends made of sulfur and carbon black prepared by dry-mixing and in composite form. The composite powder was prepared by melting the sulfur and mixing it with the porous medium made by the carbon black and then milling the composite material. For the different powders, the bulk density and packing dynamics were measured with a new and improved tapped density method (GranuPack, Granutools, Belgium) and flowability with a rotating drum method (GranuDrum, Granutools, Belgium). An increase in capacity of the produced batteries was observed to be correlated with an increase in bulk density for the composite mixing method compared to the dry-mixing method. In addition, different powder flowabilities were observed between composite and dry-mixed powder, highlighting the importance of characterizing powder properties for electrode manufacturing improvement.

280e Comparative Study on the Compression Characteristics of Coal and Biomass Powder Bed during Gas Pressurization

Shicheng Wang, Xiaolei Guo, Haifeng Liu, Haifeng Lu

Biomass substitution for fossil energy is desirable to achieve carbon emission reductions, and the storage and transportation process of biomass is governed by its compressibility. The compression characteristics of pulverized coal, rice husk and lignocellulose were compared under mechanical and gas pressurization, respectively. The experimental results showed that the descending order of relative packing density (RPD) was lignocellulose > rice husk > pulverized coal under mechanical pressurization, while an opposite trend was observed under gas pressurization. The cohesion properties were rationally correlated with the packing behavior of powder bed, and the differences in RPD order were discussed as the higher cohesion corresponding to a larger void fraction. For mechanical pressurization, biomass had higher void fraction, which was directly linked to larger RPD. On the contrary, gas permeated easily into the biomass powder bed, and the lower pressure drop decreased the compression capacity of gas pressurization. In addition, a model for predicting the RPD was proposed with an error of ± 5%.

280f Experimental Research on the Compression Behavior of Ultrafine Alumina Powders Under Gas Pressurization

Yunfei Yang, Xiaolei Guo, Shicheng Wang, Haifeng Lu, Haifeng Liu

The compression behavior of powder bed under gas pressurization plays a crucial role in the performance of pressurized powder pneumatic conveying systems. The compression behavior of ultrafine powder, especially nanoscale powder, is more complicated. In this study, a visualization method was employed to compare the compression behaviors of alumina powders with different particle sizes ranging from the microscale to the nanoscale. The experimental results revealed that the strong agglomeration led to a higher compressibility percentage of about 58 % for the nanoscale powders compared to the microscale powders with the percentage of about 44 % under gas pressurization. However, the compression degree of powders shows a non-monotonic relationship with particle size. In particular, the most compressible nanoscale powder exhibited the ability to significantly reduce the compressibility at a lower gas pressurization rate. This finding provides a new optimization method for the pressurized pneumatic conveying and feeding systems of ultrafine powder.

420b Using Digital-Twins to Relate Microscopic Parameters of DEM Simulations and Macroscopic Measurables Characterizing Powder Behavior

Geoffroy Lumay, Filip Francqui, Ben Jenkins, Aurelien Neveu, Kit Windows-Yule

DEM simulations are useful to predict the behavior of powders and granular materials inside industrial processes. For that, a digital-twin of the industrial process is designed giving access to information and measurables inaccessible in the real process. Moreover, the impact of powder characteristic modification and the impact of process parameter adaptation can be investigated. But these results are only valid if the simulation has been correctly calibrated.

Conceptually, two methods can be used for the calibration of DEM simulation parameters. The first method is a bottom-up approach based on the direct measurement of the microscopic parameters at the scale of the contacts between the grains. Even if this bottom-up method makes sense conceptually, it is often practically impossible to use this approach based on complex measurement. We propose an alternative top-bottom method based on a set of macroscopic measurements that can be used practically without difficulties to calibrate the simulation. The principle consists in measuring the powder properties in a laboratory and afterward to simulate the same process to tune the simulation parameters following an optimization process. In other words, this method is based on the use of laboratory instrument digital-twins.

From a fundamental point of view, the use of digital-twins in parallel to both laboratory characterization instruments and real processes allows to study the link between DEM simulation parameters (sliding and rolling frictions, cohesiveness, stiffness, ...) and the macroscopic behavior of the powder. In the presentation, we will illustrate this approach using the characterization instruments (GranuDrum, GranuPack, GranuHeap and GranuFlow) and typical industrial processes (recoating process in AM and capsule filling in pharma).

420a Comparison of Particle Breakage Models Accounting for Particle Size-Shape Evolution

Priscilla Hill

The pharmaceutical industry frequently produces particles by solution crystallization, and these particles are often needle shaped or acicular crystals with high aspect ratios [1]. Because the particle shape can affect the pharmaceutical's bioavailability, controlling the particle shape is essential. From a processing perspective, controlling particle shape is important because affects particle flowability and unit operations such as tableting. Therefore, it would be very useful to have accurate models that account for changes in particle shape due to mechanisms such as breakage. Despite the importance of such models to designing and controlling these processes, there are very few models available that account for particle shape evolution due to breakage. The research objective is to develop more accurate population models for breakage of high aspect ratio crystals.

Previous experimental research with needle-shaped urea crystals agitated in a dilute slurry showed the effects of breakage at two different breakage rates. Data analysis showed that the aspect ratio of the child particles produced were predominantly a function of the child particle major axis length, but not a function of the child particle minor axis length.

Novel two-dimensional (size, aspect ratio) population balance breakage models are presented and compared with each other as well as with the experimental data. A common simplifying assumption in modeling needle-shaped particle breakage is to assume binary breakage where the original parent particle breaks along a plane perpendicular to the major axis. This presentation shows that this simplified model is not sufficient to explain the experimental results.

420c Machine Learning Based Prediction of Oilfield Scale Formation Kinetics.

Glavic Tikeri, Ali Alshami

Several millions of dollars have been invested into researching the mitigation of scale formation in oilfield industries, a persistent problem that leads to decreased operational efficiency and increased maintenance costs. Here, we utilize a proactive approach, harnessing the power of machine learning to predict scale formation kinetics using data obtained from a novel continuous stirred tank laser setup. This setup offers precise control over experimental conditions and enables real time monitoring of scale formation processes at simulated oilfield conditions. This is fundamental approach, the use of machine learning models can often outperform traditional empirical or mechanistic approaches by capturing subtle trends and nonlinear relationships in the data, leading to more accurate predictions. This informs accurate inhibitor application and is a less costly approach that can be harnessed to eradicate oilfield scale formation altogether. The continuous stirred tank laser setup is used to create an extensive data set that captures the nuances of the formation of different types of scales at different replicated oilfield conditions and a machine learning algorithm is developed from it that accurately predicts the kinetics of oilfield scale formation.

420d An Innovative Approach to Studying Sifting Segregation: DEM Enabled Scale Down

Mohammadreza Ebrahimi, Stefan Pantaleev, Christopher MacTaggart, Ilgaz Akseli

Segregation is an unwanted phenomenon that can occur when handling powder mixture. It poses challenges to uniformity in product quality and, therefore, needs to be understood and mitigated [1].

During the discharge of powders from a hopper, a specific form of segregation can occur, commonly referred to as the sifting mechanism. A widely used experimental test to study sifting segregation is summarized in Shah et al.'s study [2]. In this approach, the powder mixture is discharged from one hopper to another multiple times. This transfer of material between hoppers forces the mixture to segregate if there is high segregation potential, enabling the gauging of the segregation tendency of the mixture under study. In short, the segregation tester consists of three hoppers: hopper A-1, A-2, and hopper B. Hoppers A-1 and A-2 have the same design and are low-angled, with hopper A-2 equipped with sample collectors. Hopper B is a steep-angled hopper. Initially, the powder mixture is discharged from hopper B to hopper A-1, followed by three additional transfers in the order of A1 to A2, A2 to A-1, and A-1 to A2. At the end of the segregation test, samples are collected for analysis [2].

In the pharmaceutical industry, conducting sifting segregation experiments as outlined in Shah et al.'s study [2] using the hopper size specified in that study can present significant challenges, both in terms of cost and time constraints. Access to the necessary amount of active pharmaceutical ingredient (API) for each experiment is often limited, especially during the early stages of development when the required quantities may not be available. Additionally, to reduce the uncertainty of the experiments, the test needs to be repeated at least three times, further increasing the prohibitive nature of using this segregation tester in some cases. Advanced simulation techniques, such as the Discrete Element Method (DEM), can potentially be utilized to gain insights into sifting segregation [3]. Similarly, performing full-scale DEM simulations of the sifting segregation tester can be computationally demanding, even when utilizing proposed acceleration techniques such as coarse-graining or pseudo-2-D simulation. These obstacles demonstrate the necessity for a novel and efficient approach when examining sifting segregation occurring during hopper discharge.

In this study, we present a novel approach to address these challenges by adopting a scale-down methodology inspired by the concept commonly employed when studying mixing tanks/bio-reactors. Our proposed approach involves designing scaled-down hoppers using DEM while maintaining flow regimes and stress levels experienced by the powder consistent between the actual and scaled-down hoppers. In our proposed approach, the DEM simulation of the full-scale hopper needs to be performed once to identify the flow regime and stress levels experienced by particles at the original scale. Variations of small-scale hoppers are then simulated to obtain the equivalent small-scale design of the full-scale. Once the small-scale hopper design is identified, the sifting segregation experiments can be carried out in considerably less time and utilizing fewer physical resources compared to using the original size hoppers for studying sifting segregation.

[1]. Jakubowska, Emilia, and Natalia Ciepluch. 'Blend segregation in tablets manufacturing and its effect on drug content uniformity—a review.' Pharmaceutics 13.11 (2021): 1909.

[2]. Shah, Keyur R., et al. 'Assessment of segregation potential of powder blends.' Pharmaceutical development and technology 12.5 (2007): 457-462.

[3] Hadi, Ahmed, et al. 'DEM Modelling of Segregation in Granular Materials: A Review.' KONA Powder and Particle Journal 41 (2024): 78-107.

420e A Novel Approach for Modeling Pharmaceutical Tablet Coating Using Coupled DEM-CFD Modeling

Lucilla Coelho de Almeida, Tim Hoogenraad, Vinicius Daroz, Rahul Bharadwaj, Ariel Muliadi

Oral drug delivery via tablets is highly favored across numerous therapeutic areas for its convenience, precise dosing, stability, and patient adherence. Tablets often receive a thin film coating on their surface. In most cases, the coating is applied simply for cosmetic reasons, e.g. branding, blinding, etc. A small subset of tablet formulations, however, require a functional coating. An example of functional coatings is the one applied to prevent drug dissolution in the stomach and instead release it in the small intestine (i.e. enteric coating). This coating is essential for protecting the drug from gastric acid degradation or inactivation, which is often a requisite for oral delivery of macromolecules.

The coating process is performed on a rotating perforated cylindrical pan with conical ends, often with baffles designed to tumble the tablets therein. The coating formulation is then sprayed onto a small area of the tablet bed via an atomizer. The spray droplets that deposit onto the tablet surface as it passes through the spray zone are then dried to form a coating film by the warm air that is passed through the pan's perforations. The process is continued until an established endpoint, typically an average tablet weight gain, is achieved. For nonfunctional (i.e. cosmetics) coating, an average weight gain of 4-5% is usually targeted. For enteric coating, the target weight gain can be as large as 15-20%.

Tablet weight gain is typically assessed by first taking the average weight of 10-20 (uncoated) tablets before starting the coating process. Once the coating begins, the same number of tablets as that initially measured are drawn from the pan. Their weight is then measured. This process is repeated at a fixed time interval until the target weight gain is achieved. In the case of enteric coating, the weight of individual tablets representing a small portion of the batch may be measured as well to ensure that the inter-tablet coating uniformity (i.e. the amount of coating deposited between one tablet and the next) is adequate. While this approach to quality control may seem primitive, it is frequently deemed sufficient because measuring the weight of all tablets in a batch is practically impossible, especially when the batch size is large (i.e. hundreds of kilograms). Accordingly, an alternate method that better quantifies the process outcomes, such as numerical methods that incorporate all key process physics and thus can predict the inter-tablet coating uniformity more accurately than the above experimental approach, can be extremely valuable for process development and validation.

This study presents the development of a novel DEM model using Ansys Rocky to predict inter-tablet coating uniformity. The model incorporates the actual tablet shape (instead of using a "glued spheres" approach). A hysteretic linear spring model is used to compute normal contact forces, and a linear spring Coulomb limit model calculates tangential forces. Spray droplets are modeled as DEM particles and tracked within the DEM framework, allowing the local spray properties (i.e. droplet concentrations at a given location, their velocities, and size distributions) to be captured and thus simulating the spray dynamics in the coating process. Contact detection between spray particles is omitted to save computational resources. Additionally, upon collision with a tablet, a droplet is immediately removed from the simulation, and its mass is added to the liquid film mass on that tablet. In addition, the size of each tablet is also increased in the simulation to account for the increasing coating mass deposited on them, which, in the case of enteric coating, can be significant when compared to the uncoated tablet weight itself.

To better model the spray-tablet interactions, the mass, momentum, and energy transfer between the spray droplets and the underlying flow field are also modeled by coupling Ansys Rocky and Ansys Fluent solvers. A droplet evaporation model is implemented by computing the instantaneous local evaporation rate based on the difference in water vapor concentration between the particle surface and its surroundings. As the water evaporates, the remaining solute fraction is integrated into the coating deposit on the tablet surface from the previous time step. As explained above, this additional mass (and volume) is then used to increase the mass and size of the tablets in the simulation. Finally, a liquid bridge model is also incorporated to account for liquid transfer between tablets that may occur during tablet-tablet collisions, as well as the effect of viscous lubrication and capillary forces on the particle bed dynamics.

We demonstrate the effects of adjusting process parameters such as droplet size, spray rate, drum speed, and temperature on process outcomes such as the time to reach the coating endpoint, inter-tablet coating uniformity, and tablet bed dynamics. Model results are also compared to experimental data from bench-scale tablet coating runs.

420f Predicting the Permeability of Porous Media Using Graph Neural Networks

Jack I'Anson, Mark Simmons, E. Hugh Stitt, Robert Gallen

Introduction

Understanding of the properties of porous media is beneficial for a range of important applications including fuel cells [1] and automotive exhaust treatment [2]. However, determining properties such as permeability can be difficult, both experimentally and via simulation. Typical approaches for numerically determining such properties involve some type of pore-scale simulation. Two common examples include traditional Computational Fluid Dynamics (CFD) [2] and the lattice Boltzmann method (LBM) [3]. These methods require significant computational cost, often requiring the use of high-performance computing (HPC) [4]. Alternatively, properties can be determined via experimentation, for example by driving flow through porous media and measuring pressure drop [5], or extracting pore network models from mercury porosimetry data [6]. These methods are useful for validating simulations, but require complex experimental set-ups, and more data can be gathered from simulations.

Various data-driven methods have been used for predicting porous media properties. A common method is to use a convolutional neural network (CNN), trained on images [7]. Using 2D image data is more computationally efficient, but often characteristics of 2D porous media, such as dead-end pores, are not present in 3D porous media [8]. However, 3D CNNs are computationally expensive and it may prove difficult when dealing with porous media samples that are representative of the structures from which they were extracted when trying to scale up the CNN. An alternative is to train a graph neural network (GNN) based on pore networks. Cai et al. [9] trained a graph isomorphism network, a type of GNN, to make predictions of porous media properties using Morse graphs extracted from porous media images.

While several methods have been used to predict the properties of porous media, it is important that they are scalable enough to handle representative samples. This work aims to solve this problem by training a GNN on representative pore networks extracted from the samples, coupled with some structure properties such as porosity, capable of generalised to new, unseen structures. The contents of this abstract and presentation are discussed in more detail by I'Anson et al. [10] (to be submitted for publication).

Methods

A summary of the development framework for producing a GNN to predict the permeability of porous media is shown in Figure 1.

Dataset Preparation

Training deep learning models to predict porous media properties typically requires thousands of samples [7]. However, creating a dataset based on real porous media samples is difficult because imaging techniques, such as X-ray tomography, are required. To overcome this, a dataset comprising randomly generated porous media was produced using fractal noise generated from PoreSpy. This allowed for realistic, binary structures with varying properties to be generated quickly when compared to the preparation of an experimental dataset with similar variance. Pore networks were extracted from the structures, providing information about the pore size and connectivity, and will be used to train the GNN.

The LBM was used to obtain the ground truth for the permeability of the porous media samples. The LBM was chosen because of its applicability to domains with complex geometries, such as porous media [11]. Flow was simulated through each structure and the resulting velocity and density fields were used to compute the permeability using Darcy's law.

Performing the LBM simulations for the thousands of samples needed to train the GNN is infeasible due to the required computational costs. Therefore, a set of 250 'parent' structures voxels in size were generated; the density and velocity fields were obtained for this set using the LBM and pore networks were extracted from each structure. Thousands of smaller, overlapping samples were extracted by sampling the voxel structure, the pore network, and the simulation results. A representative elementary volume (REV) scale analysis was performed on the structures based on the number of pores in the pore network, rather than the number of voxels in the structure. This leads to a scale-free metric for REV-scale, otherwise structures of varying pore size would require different REV-scale values. This led to a training dataset of around 250,000 representative samples which was used to train the GNN. Two additional independent datasets were generated from new 'parent' structures; one for validation during the hyperparameter selection and one for the final model testing.

Graph Neural Network

The GNN was developed and trained using the Python deep learning library PyTorch, which enables large models to be trained quickly. The GNN requires the pore network to be encoded into a feature matrix, which contains pore features such as pore volume and surface area, and the adjacency matrix which encodes each of the connections in the pore network. The GNN is split into two main sections: a set of message passing layers, and a set of fully connected layers. The message passing layers learn node embeddings from the feature matrix and the adjacency matrix. Following the message passing layers, the updated feature matrix is then flattened and passed to the fully connected layers, resulting in a single node prediction of the natural log of the permeability. The sample permeabilities span nearly five orders of magnitude, therefore predicting the log permeability allows samples with a large permeability to be weighted equally to samples with a low permeability during model training. The optimal hyperparameters for the GNN were found using the optimisation framework Optuna, which allowed the large hyperparameter space to be explored efficiently. The optimal model was then tested on a new set of structures and compared to a benchmark method, the Carman-Kozeny equation.

Results

Dataset Preparation

The LBM simulations were performed using a C++ LBM library, Palabos using a HPC cluster. The simulations took several hours to complete. Figure 2a and 2b shows a 2D slice of the density field in the direction of flow and the 3D velocity field obtained from the simulations. The structures had permeability values spanning several orders of magnitude, depending on porosity and pore size within the structures.

The maximum sized pore network in the dataset had over 8000 pores, but it was found that any sample with more than around 5000 pores in the pore network was an outlier, and therefore was removed from the dataset. The REV-scale analysis indicated that each sample structure needed a minimum of around 1500 pores in the pore network to be representative. This number is independent of the structure voxel count, porosity and permeability, allowing for a more consistent representative sample size for the GNN input. Most of the samples were within the range of 1500-5000 pores, leaving a large training dataset to be used for the GNN.

GNN Performance

The GNN with the optimal hyperparameters was trained for a total of 160 epochs on the training dataset. This was because the model performance on the testing dataset was not improving after 160 epochs, indicating the model was starting to overfit, lowering model generalisability. Figures 3a and 3b show the error plot during the training process and the parity plot showing a comparison of the LBM permeability and the predicted permeability for the training and testing datasets. To assess the performance of the GNN, the permeability was estimated using the Carman-Kozeny equation based on the porosity and the specific solid surface area. Figure 4 shows a comparison of the LBM to both the GNN and the Carman-Kozeny predictions. The GNN had a mean squared error (MSE) of 0.0373, compared to the Carman-Kozeny equation which had an MSE of 0.432 vs the LBM prediction.

Conclusions

In this work, a GNN was trained to predict the permeability of randomly generated, representative porous media samples. This method allows for rapid (near-real time) predictions of the permeability from 3D structure data. The REV-scale analysis indicated that the pore networks used to train the GNN should have at least 1500 pores, meaning structures of any size (in terms of voxels) could be passed to the GNN providing they contain enough pores. This is an important requirement for the model, as failure to be able to handle representative samples will lead to a less accurate model. The GNN was able to make near real-time and accurate predictions with a significantly lower MSE compared to the Carmen-Kozeny predictions when compared to the LBM simulations, which typically require hours to complete.

References

[1] Z. Niu et al., Energy Environ. Sci. **14**, 2549 (2021).

[2] P. Kočí et al., Catal. Today **320**, 165 (2019).

[3] A. Eshghinejadfard et al., Int. J. Heat Fluid Flow **62**, 93 (2016).

[4] T. Yasuda et al., Chem. Eng. J. **420**, 130069 (2021).

[5] A. Wagner et al., Transp. Porous Media **138**, 1 (2021).

[6] Q. Xiong, T. G. Baychev, and A. P. Jivkov, J. Contam. Hydrol. **192**, 101 (2016).

[7] H. Wang et al., Energy AI **2**, 100035 (2020).

[8] H. J. Haugen and S. Bertoldi, in *Characterization of Polymeric Biomaterials* (Elsevier, 2017), pp. 21–53.

[9] C. Cai et al., Int. J. Multiscale Comput. Eng. (2021).

[10] J. I'Anson et al., (2024) (to be submitted).

[11] T. Krüger et al., in *The Lattice Boltzmann Method: Principles and Practice* (Springer International Publishing, Cham, 2017), pp. 31–58.

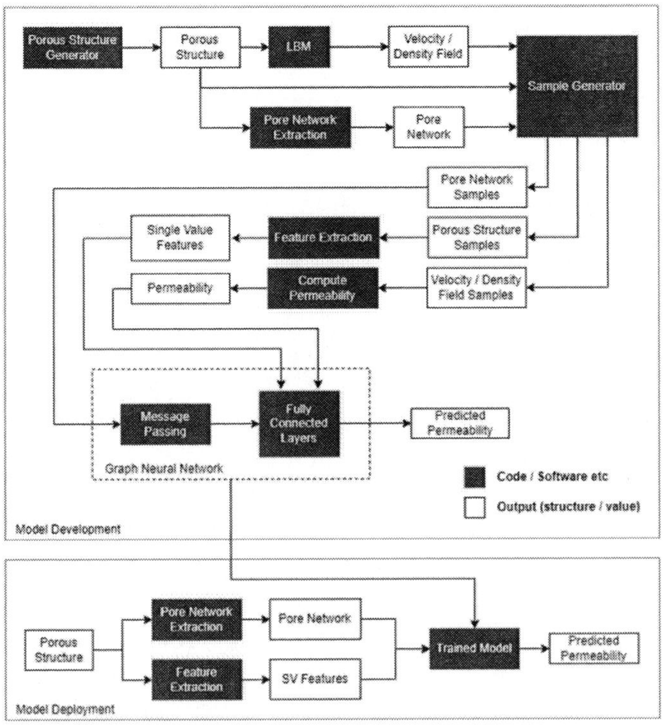

Figure 1: A flow diagram showing the development of the GNN and how it would be deployed

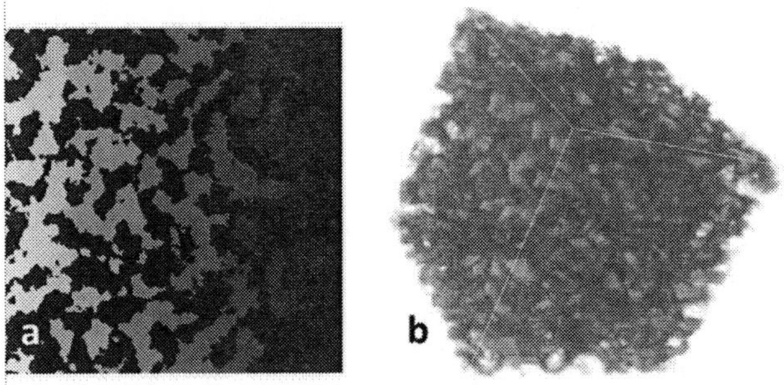

Figure 2: a) a 2D slice taken from the pressure gradient for flow to the right with higher pressure in yellow, and b) a 3D velocity field

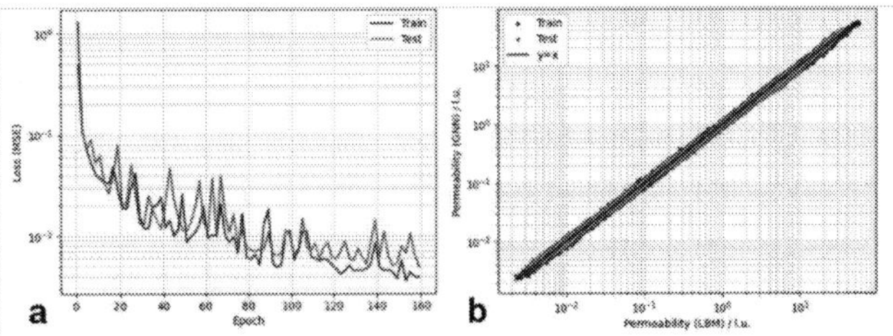

Figure 3: a) The training and testing MSE per epoch of the training process and b) the parity plot for the final chosen model after 160 epochs.

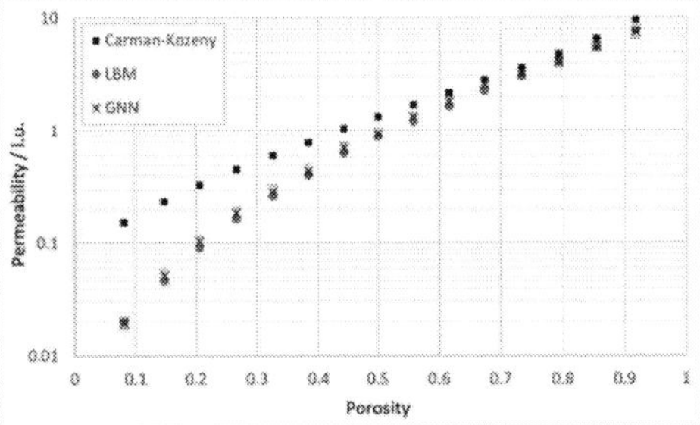

Figure 4: A comparison of the permeability obtained from the GNN, the LBM and the Carman-Kozeny equation for structures of varying porosity.

420g Spray Drying for Uniform Sized Particle Formation: A Multiscale Modeling and Simulation Approach

Jie Xiao

Spray drying has been widely used in food, pharmaceutical and energy industries for powder production. Although conventional spray dryers can produce huge amounts of powder products without complicated operations, they can hardly be used for functional powders, whose complex structures need to be precisely controlled for quality assurance. Mono-disperse droplet spray dryer (MDSD) is a new technology that can offer uniform sized particles with almost identical morphology. It is a promising lab-scale platform for exploring functional powders. However, its production efficiency and quality control capability need to be improved.

In this talk, we will share with the audience our past and on-going efforts on developing multiscale models for MDSD to realize virtual manufacturing of functional powder products. We have been pursuing multiscale models for the integrated system of the spray drying process and the particle product. At the dryer scale (macro), a comprehensive CFD model has been developed to take care of the multi-stage drying process. Coupling the level set model with a solid mechanics model of a piezo-ceramic nozzle allowed us to computationally explore mono-disperse droplet formation through atomization. A unique discrete phase model (DPM) was developed to track evaporating droplets flying through chambers with distinctively different sizes, i.e., the small dispersion chamber followed by the big drying chamber. For the first time, droplets in MDSD can be tracked from generation to the end of the drying stage. At the single droplet scale (meso), a liquid-gas two phase flow model coupled with the moving mesh ALE method was developed to simulate the droplet drying process. The ability of quantifying the local evaporation rate and characterizing crust formation enabled us to track the shape evolution of a shrinkage droplet. At the droplets' surface layer scale (micro), a coarse-grained lattice Monte Carlo (LMC) approach is being developed to simulate evaporation induced self-assembly of solute molecules. Evolution of the nanoscale surface morphology and composition under different drying conditions could then be systematically investigated. By resorting to this multiscale simulation platform, a number of new designs and optimized operations have been successfully identified, which can lead to improved production rate and reduced cost without sacrificing powder quality. In addition to the above listed advances, we would also like to discuss with experts in the audience the key challenges faced by us when developing individual-scaled models with a hope to seek collaborations.

652a Spherical Agglomeration of Battery Materials: Mechanistic Understanding and Hollow Agglomerate Formation

Rachel Smith, Jediah Capindale, Denis Cumming

Battery electrode microstructure is known to strongly influence both capacity and rate capability for lithium ion batteries. High tortuosity and disrupted conductivity networks limit ion and electron transport in batteries, and therefore control of electrode microstructure can improve high-performance electrodes.

Spherical agglomeration is an in-suspension agglomeration technique, which uses the addition of immiscible fluids to generate agglomerate structures with desired particle attributes. In this study, the spherical agglomeration technique is manipulated to produce hollow co-agglomerated particulate structures for application in battery electrode manufacture. In particular, the rate limiting particle immersion mechanism is exploited, which allows for the tailoring of particle structure.

High value characterisation techniques such as SEM, XRCT and particle size measurements are used to examine agglomerate structure, and to provide insight into the controlling mechanisms of formation. A thermodynamic analysis found that interfacial energy gradient is seen to be pivotal for particle transfer between immiscible fluids, and thus successful spherical agglomerate systems. A regime map is proposed which aids in the generation of novel agglomerate structures, based on a modified capillary number describing the ratio of viscous dissipation to the work of particle transfer.

Agglomeration of battery electrode materials has many potential promising benefits including improved homogeneity, reduced carbon migration and improved particle handling, whilst the process can operate at ambient conditions. The results of this research offer a potential route for the development of intentionally designed high-performance battery electrodes, and demonstrates the ability to create hollow particulate structures using advanced manufacturing techniques. The process also shows great promise for other industries such as pharmaceuticals and agrichemicals.

652b Interface Engineering for Optimizing Performance of the Ultra-High Nickel Cathode in All-Solid-State Batteries

Wenjin Li, Kaiyuan Deng, Cheng Liu, Puxi An, Shaoyang Song, Lei Yao, Guangliang Liu

All-solid-state lithium batteries offer the advantages of high energy density and superior safety, making them an ideal choice for next-generation battery technology. In all-solid-state lithium batteries, ultra-high nickel cathode materials can achieve a remarkable energy density of 500Wh/kg. However, during prolonged charge-discharge cycles, these batteries face challenges such as micro-cracks within particles of ultra-high nickel cathode materials, failure in solid-solid contact at interfaces, and surface chemical side reactions. To address these issues, a surface engineering design strategy has been employed to construct various nano-oxide coatings and oxygen-vacancy-rich lithium oxide coatings on the surface of ultra-high nickel cathode particles. This approach aims to enhance the interface stability of ultra-high nickel cathodes by reducing the internal micro-cracks within cathode particles and minimizing the surface side reactions that occur between cathode and solid electrolyte particles during battery cycling. Finally, both the long cycle life and fast charging performance of all-solid-state batteries are improved.

652c A Two-Step Size Reduction Method for Fine Milling of Electronic Waste to Improve Recycling

Yidong Xia, Noah Berglund, Jordan Klinger, Tiasha Bhattacharjee, Vicki Thompson, Zachary Diermyer, Jiaoyan Li

Electronic waste (e-waste) recycling is important for waste treatment and recovery of valuable materials, as it can reduce the extraction of virgin ores and extend the life of those resources. Hydrometallurgical recovery of valuable metals from printed circuit boards (PCBs) is widely used and consists of at least two main unit operations of leaching and recovery. During leaching, a smaller particle size results in a greater dissolution of valuable metals, due to higher surface area and higher proportion of metals exposed to the leaching fluid. Conventional size reduction process involves shredding and hammer milling, for which industries reported as much as over 60% of valuable elements is lost through those unit operations in output streams (milling dusts), resulting in an overall reduction of the recycling efficiency and implying the need for improvement. A novel two-step mechanical size reduction method for pretreatment of e-waste materials is being developed for generating high-efficiency leaching-ready e-waste feedstock. This method involves a shredder mill and a cryogenic mill at Idaho National Laboratory's (INL's) Biomass Feedstock National User Facility (BFNUF). In a conventional two-step milling process, the hammer mill was found prone to generating sparks and presenting fire hazard during operation. In addition, loss of mass in dust and fines is not well mitigated. To re-risk the operation and improve efficiency, a new cryogenic mill is used instead and re-configured for reducing mass loss in processing e-waste. The mill chamber is filled with liquid nitrogen during operation, which significantly increased the brittleness of the particles and enhanced fragmentation. Both the discharge and vent valves are specially configured to allow stable ventilation of nitrogen gas and retention of fine particles. The size of the resultant fine particles ranges from sub-100 microns up to a few millimeters. Mostly importantly, the mass loss rate is below 5%, a huge improvement over the current industrial level.

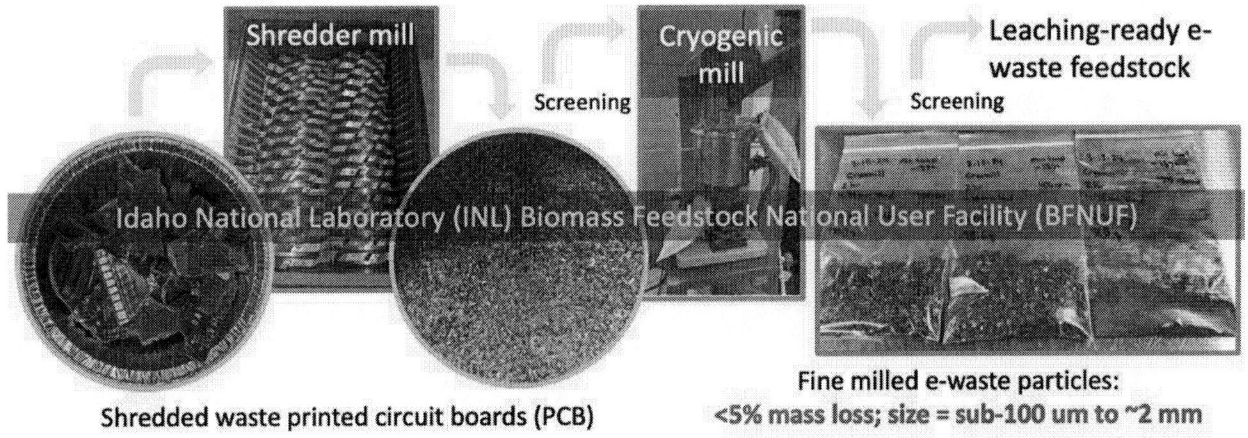

652d Recycled Polyurethane Polyol Dispersion with Improved Stability

En Wang, Craig Gorin, Kathryn Grzesiak, Hari Katepalli, Daniel Dermody, Hans Kramer, Kaoru Aou

The majority of post-consumer polyurethane waste today is sent to a landfill or incinerated. To enable a circular economy, Dow Polyurethanes has recently launched the RENUVA™ Mattress Recycling initiative. In the novel hybrid glycolysis/acidolysis process, polyurethane foam from mattresses will be recycled into polyols that can be reused in a new polyurethane foam formulation. These recycled polyurethane polyol products have particles sedimentation occurring over time. This work comes up with stabilizer additives to improve recycled polyurethane polyol product stability. The stabilization solutions do not lead to a significant increase in the viscosity, nor interfere with a polyurethane foaming process that makes use of the product.

652e Leveraging Discrete Element Method Simulations for Studying Mechanochemical Depolymerization of Polystyrene Waste

Jacob Sweet, Elisavet Anglou, Yuchen Chang, Carsten Sievers, Fani Boukouvala

Mechanochemical depolymerization is a novel chemical recycling route that seeks to convert solid waste plastics into their constituent monomers through the application of mechanical forces and energy [1]. Polymers are loaded into mechanically agitated ball mills where they experience local high-energy collisions with grinding media. Subsequently, those events cause successive breakage of the chemical bonds of the polymer chains. The high dissipated energies and local temperatures experienced obviate the need for costly heating and large solvent volumes associated with traditional depolymerization methods thereby reducing the total environmental footprint [2, 3].

Understanding the mechanisms of depolymerization within mechanochemical recycling processes is critical to improving energetic efficiency, informing industrial scale-up studies and ultimately optimizing the mechanochemical process [4, 5]. This presentation focuses on the investigation of the kinetic rates and mechanisms of depolymerizing polystyrene (PS) waste in a lab-scale vibratory ball mill [6]. The inside of the mill contains many high-velocity particles and multiple grinding media, so the in-situ examination of particle kinetics and heat dissipation is experimentally infeasible [7]. To estimate the particle motions and energetics within the mill, a Discrete Element Method (DEM) simulation was developed and validated. The outcome of those simulations is a detailed representation (e.g., every tenth of a millisecond) of the movement and forces of both grinding media and polymer particles within the moving mill. Building off extensive particle comminution literature that leverage DEM outputs such as collision energy, heat dissipation, and compressive forces to predict particle breakage kinetics, we develop DEM informatics that predict depolymerization reactions [8, 9]. We leverage machine learning methods to analyze large, simulated datasets of high dimensionality to consider and compare potential indicators of depolymerization. Furthermore, by distinguishing particle-particle, particle-ball, and other collision types, as well as by tracking the successive collisions experienced by individual plastic particles, we use the DEM simulations to gain insight into the depolymerization reaction mechanism.

Finally, results from the DEM simulations are subsequently used as inputs to a population balance model (PBM) to form a hybrid model from which we can elucidate the temporal evolution of the PS molecular weight distribution (MWD) during depolymerization. The hybrid PBM retains the structure of traditional milling PBMs, but it is parameterized by physically relevant outputs of DEM simulation that predict depolymerization rates. The hybrid PBM is further informed by diverse experimental data and utilizes recently advanced machine learning methods to better predict MWDs at unseen conditions and to better understand kinetic patterns within the mill. Gaining a better understanding of the kinetics within the mechanochemical depolymerization process and capturing the dynamic evolution of MWDs throughout depolymerization are necessary tools for the effective optimization and industrial scaling of the mechanochemical recycling solution.

Citations

[1] I. Vollmer *et al.*, "Beyond Mechanical Recycling: Giving New Life to Plastic Waste," *Angew. Chem. Int. Ed.*, vol. 59, no. 36, pp. 15402–15423, 2020, doi: 10.1002/anie.201915651.

[2] A. W. Tricker, G. Samaras, K. L. Hebisch, M. J. Realff, and C. Sievers, "Hot spot generation, reactivity, and decay in mechanochemical reactors," *Chem. Eng. J.*, vol. 382, p. 122954, Feb. 2020,

doi: 10.1016/j.cej.2019.122954.

[3] A. W. Tricker *et al.*, "Stages and Kinetics of Mechanochemical Depolymerization of Poly(ethylene terephthalate) with Sodium Hydroxide," *ACS Sustain. Chem. Eng.*, vol. 10, no. 34, pp. 11338–11347, Aug. 2022, doi: 10.1021/acssuschemeng.2c03376.

[4] E. Anglou, Y. Chang, A. Ganesan, S. Nair, C. Sievers, and F. Boukouvala, "Discrete Element Simulation and Economics of Mechanochemical Grinding of Plastic Waste at an Industrial Scale," in *Computer Aided Chemical Engineering*, vol. 52, A. C. Kokossis, M. C. Georgiadis, and E. Pistikopoulos, Eds., in 33 European Symposium on Computer Aided Process Engineering, vol. 52. , Elsevier, 2023, pp. 2405–2410. doi: 10.1016/B978-0-443-15274-0.50382-6.

[5] E. Anglou *et al.*, "Process development and techno-economic analysis for mechanochemical recycling of poly(ethylene terephthalate)," *Chem. Eng. J.*, vol. 481, p. 148278, Feb. 2024, doi: 10.1016/j.cej.2023.148278.

[6] Y. Chang *et al.*, "Kinetic Phenomena in Mechanochemical Depolymerization of Poly(styrene)," *ACS Sustain. Chem. Eng.*, vol. 12, no. 1, pp. 178–191, Dec. 2023, doi: 10.1021/acssuschemeng.3c05296.

[7] F. Delogu and L. Takacs, "Information on the mechanism of mechanochemical reaction from detailed studies of the reaction kinetics," *J. Mater. Sci.*, vol. 53, no. 19, pp. 13331–13342, Oct. 2018, doi: 10.1007/s10853-018-2090-1.

[8] M. H. Wang, R. Y. Yang, and A. B. Yu, "DEM investigation of energy distribution and particle breakage in tumbling ball mills," *Powder Technol.*, vol. 223, pp. 83–91, Jun. 2012, doi: 10.1016/j.powtec.2011.07.024.

[9] A. Datta and R. K. Rajamani, "A direct approach of modeling batch grinding in ball mills using population balance principles and impact energy distribution," *Int. J. Miner. Process.*, vol. 64, no. 4, pp. 181–200, Apr. 2002, doi: 10.1016/S0301-7516(01)00044-8.

652f Investigation of Amine Silane Film Formation Via Particle Molecular Layer Deposition for CO_2 Capture

Hailey Loehde-Woolard, Bergen Evans, Michael Zaza, Katarina Odak, Kent J. Warren, Alan W. Weimer

Particle Molecular Layer Deposition (MLD), closely related to Atomic Layer Deposition (ALD), involves a series of alternating gas phase reactions on a particle surface. Our work studies particle MLD as a method of generating aminosilane films on a variety of solid supports to create CO_2 capture materials. In addition, we describe the construction and validation the custom particle ALD/MLD system used in this work.

In the past few decades, solid adsorbent materials have been a focus of research as an alternative to traditional liquid amine absorbent systems which require substantial regeneration energy and cost. Supported amine adsorbents have gained much interest as it maintains the traditional acid-base chemistry, operates under ambient conditions, and is tolerant of humidity. However, the current liquid-based functionalization techniques have limitations; grafted materials tend to have lower adsorption capacities and impregnated materials tend to have poorer regeneration abilities. (Unveren, 2017) In addition, liquid phase grafting is heavily dependent on numerous process conditions and produces amorphous 3-dimensional films. (Yadav 2013) Vapor deposition of aminosilanes has been demonstrated to produce reliable mono- and sub-monolayer films, which provide greater surface control but has not been studied in depth for CO_2 capture applications. (Yadav 2013, Rozyyev 2021) Vapor deposition also differs from MLD, as vapor deposition only exposes the surface to one reactant whereas MLD regenerates reactive groups on the surface through a series of reactions in each cycle.

Here, we use MLD to deposit a covalently bonded aminopropylsiloxane network on particle supports. A custom vibrating fluidized bed particle ALD reactor was constructed in house and validated through fluidization experiments and deposition of aluminum oxide as a benchmark chemistry. Following validation, the system was employed to investigate the MLD chemistry of (3-aminopropyl)triethoxysilane (mono-amine) and water. Sorbent materials were generated according to design of experiments that focused on factors related to film formation such as the number of cycles and deposition temperature. For comparison, materials were also generated through traditional liquid phase grafting and vapor phase deposition. Characterization of the sorbents included BET surface area analysis, LECO elemental analysis, X-ray photoelectron spectroscopy (XPS), and thermogravimetric analysis (TGA). The adsorption capacity was shown to be affected by both the number of MLD cycles and the temperature of deposition. It was previously found by Rozyyev et al. that for vapor deposition, lower deposition temperature led to greater amine content. Interestingly, MLD materials produced at higher temperatures lead to higher adsorption capacity.

Ünveren, E. E.; Monkul, B. Ö.; Sarıoğlan, Ş.; Karademir, N.; Alper, E. Solid amine sorbents for CO 2 capture by chemical adsorption: A review. Petroleum 2017, 3 (1), 37

Weimer, A.W. Particle atomic layer deposition. J. Nanopart. Res. 2019, 21:9

Yadav, A. R., Sriram, R., Carter, J. A., Miller, B. L. Comparative study of solution-phase and vapor-phase deposition of aminosilanes on silicon dioxide surfaces. Materials Science and Engineering C, 2014, 35, 283-290

Rozyyev, V., Murphy, J. G., Barry, E., Mane, A. U., Sibener, S. J., Elam, J.W. Vapor-phase grafting of a model aminosilane compound to Al_2O_3, ZnO, and TiO_2 surfaces prepared by atomic layer deposition. Applied Surface Science, 2021. 562, 149996

488a Fine Particle Systems – a Personal View on Multidimensional Particle Characterization

Wolfgang Peukert

Reg Davies was an outstanding pioneer of particle science and technology. His PARSAT team at Dupont belonged to the best and most influential groups in industry worldwide. A long-term vision in particle science and technology is the targeted design of particulate products including their surfaces, as mentioned by Reg in his outstanding review *Particle Science and Technology – a view at the millennium*. Starting from his views on the future of fine particle systems I will describe progress in systematic approaches to rigorous optimization of particle formation, formulation and characterization based on predictive structure-property and process-structure functions. As Reg was also a pioneer in particle characterization, I will discuss recent developments in multidimensional characterization of particles with respect to size, shape, composition and surface properties for a range of different material systems. Recent developments in multiwavelength analytical ultracentrifugation and nanoparticle chromatography in the liquid phase as well as electrospray-assisted mobility-classified mass spectrometry in the gas phase allow nowadays the precise and amazingly accurate multidimensional and multimodal characterization of fine particles.

Funding by DFG of the Collaborative Research Center 1411 "Design of Particulate Products" is greatfully acknowledged (https://www.crc1411.research.fau.eu/).

488b Particle Formation as the Foundation of Processes to Produce Particulate Products

Daniel A. Green, Lotfi Derdour

Reg Davies recognized the importance of viewing the overall production of particulate products as the sum of individual operations from particle formation, through downstream processing, operations to tailor particle properties, particle conveying, and finally packaging. This talk will focus on particle formation, the process step that creates particles in their most basic form. Significant efficiency and even improvement in particulate product performance are gained when particle formation is performed well, with the process establishing desired particle properties at the very start of the overall process. Designing high-performing particle formation processes requires an understanding of the mechanisms of particle formation science and how fundamental particle properties affect performance in use. Of course, it is not always possible to form particles with all the properties desired of the final product. In this case, we recognize the benefit of meshing particle formation with downstream processing that manipulates particles to provide those desired properties. I will review how fundamental particle characteristics established at formation affect particle performance properties, and how these characteristics are changed and manipulated during the overall process. Examples are drawn from multiple industries.

488c 70 Years of Research and Application in Solid-Liquid Mixing

Richard Grenville

Suspension of solid particles in liquids is an important mixing operation in the process industries. These include crystallization and precipitation where particles are formed, dissolving of solids to create solutions and reactions where solid particles are catalysts. The optimum agitator speed for mass transfer between the particles and liquid is called the just suspension speed and was defined by Zwietering (Chem. Eng. Sci., 8, 244-253, 1958) as the speed required to ensure that no particle is stationary on the base of the vessel for longer than 1-2 seconds. Zwietering produced an empirical correlation, based on dimensional analysis, which correlated his data and has been used for next 50 years to design agitators for these applications. The correlation was recommended for use in the Handbook of Industrial Mixing (ch.10, John Wiley & Sons Inc., Hoboken NJ, 2004).

In the late 1970's and early 1980's physical models were proposed by Baldi et al. (Chem. Eng. Sci., 33, 21-25, 1978) and Davies (Chem. Engng. Proc. 20, 175-181, 1986) which related the properties of the turbulent eddies generated in the flow from the impeller to the negative buoyancy of the particles. Grenville et al. (ChERD, 100, 282-291, 2015) applied the Baldi / Davies physical modeling approach to a large data set measured by the Fluid Mixing Processes consortium in the UK. Measurements were made at three scales with liquid viscosities ranging from 1 to 2000 cP. The model captures the variability in the data very well taking into account scale, impeller type and geometry and physical properties of the particles and liquid.

Zwietering's correlation predicts that, on scale-up, the power input by the impeller per mass of slurry will decrease while Grenville et al. show that it is constant, and this is an important conclusion for designers of industrial scale equipment. Giacomelli et al. (Flow. Turb. Comb., 105, 31-62, 2020; AIChEJ., 67, 2, 2021; ChERD., 180, 318-332, 2022;ChERD., 190, 793-813, 2023) have developed a CFD model to confirm the underlying assumptions of the Baldi / Davies model and extended the applicability of the Grenville et al. correlation to vessels with other, industrially relevant geometries.

In this presentation we will review the history of this important area and discuss recent developments in research.

488d Solid-Liquid Separation – An Overview and the Challenges

Karsten Keller

Solid-liquid separation is a critical process in various industries, including food, pharmaceuticals, and chemicals. The particle size significantly influences the separation technique used. For larger particles,
methods like filtration, centrifugation, or gravity settling might be sufficient. However, when dealing with nanosized particles, the separation becomes more challenging due to their particle / particle interactions. Membranes are used frequently for very small particles.
In addition to the physical properties of the particles, other factors like the concentration of the particles
and the nature of the liquid medium also play a significant role in determining the most suitable solid-liquid separation method. A comprehensive understanding of these factors is essential for optimizing the separation process and ensuring efficient and cost-effective operations in the respective industries.
The presentation will provide an overview of solid-liquid separation in the industry and the challenges to
find cost effective solutions.

488e Relevance of Particle Technology in Bio Based Manufacturing.

Erik Gommeren

A proper understanding of particle technology is important during recovery, purification and formulation of fermentation based products. In this talk I review common particle related challenges we face in Industrial Biotechnology during our manufacturing steps and how particle properties impact performance of our end products. The types of products include enzymes, probiotics, food ingredients and biopolymers.

488f Relevance of Solids Processing Expertise to the Chemical Industry – Past, Present and Future - a Perspective from Dow Chemical

Shrikant Dhodapkar, Jorg Theuerkauf

This talk is focusing on the historic perspective of solids handling at Dow. Specially, design and scale up of unit operations will be discussed and how understanding the processes fundamentals had to be developed. Due to the nature of solids, it was important to conduct meaningful laboratory experiments coupled with fundamentals to successfully scale up or trouble shoot processes on a plant scale. However, today there are a variety of tools available which can augment or even replace some of the experiments necessary to gain the understanding. We are discussing a view of current tools for process analysis and design, the challenges and use of computational tools for individual unit operations using the discrete element method, finite element analysis and computational fluid dynamics as well as process simulations using flow sheeting tools. Besides the computational tools we will discuss advances in laboratory and process instrumentation, including a wide range of particle and powder characterization tools and methods which enable process intensification. Further, we will discuss the needs for more powerful computational tools, how the aspect of sustainability is considered and the need for rapid innovation in the field.

488g Effect of Particle Size, Shape and Structure on the Moisture SorptionBehaviors of Milk Powders

Keisha Wilson McDowell, Steven F. Krakowski

Skim milk powders were produced via spray-drying and freeze-drying processes, and water-based moisture sorption isotherms were determined at various temperatures using a high-resolution dynamic vapor sorption method. The isotherms of the laboratory-dried powders were compared with isotherms of commercially granulated skim milk and showed that the absorption behaviors varied depending on the particle size, shape, and structure of the milk powders. Two- and three-parameter moisture sorption equations were used to model the moisture sorption behaviors and identify moisture-induced transition events. The models showed good fit to the data, and differences in the fitting parameters could be related to the structures generated via the dehydration and/or granulation process used. These data are useful in determining optimal powder handling and storage stability conditions and in packaging selections for dried food products exhibiting differing particulate properties.

488h Analysis of Concentrated Solar Energy Storage Using Packed and Fluidized Bed Systems

Hamid Arastoopour, Zeyuan Gao, Javad Abbasian

Concentrated solar energy (CSE) is capable of providing the energy needed for the majority of chemical, pharmaceutical, and biological processes. One of the challenges in using CSE is the development of innovative fluids or mixtures of fluid and particle systems to efficiently adsorb concentrated solar radiation and transfer the stored energy for utilizations. In this presentation a radiative heat transfer model was developed and linked to computational fluid dynamics (CFD) model to analyze CSE absorption by both packed and fluidized bed of silicon carbide (SiC). The results of our simulation agreed well with temperature measurement by infrared camera on a directly irradiated packed and fluidized bed solar absorbers experimental data in the literature. Our simulation showed that the fluidized bed reduces the surface temperature by convective energy transfer for the bubbling fluidized bed due to the mixing created by bubbles. The lower temperature on the surface significantly decreases the radiative energy loss from the surface to the environment, which results in larger energy absorption by the bubbling fluidized bed compared with similar packed bed system. Furthermore, using a wider incident radiative energy flux distribution would increase the energy absorption by both packed and fluid bed systems. Our numerical simulations showed that a packed bed system with a novel design of an optical quartz tube, where the radiative energy from the bed surface meets the surface of the optical tube at the center of the packed bed, results in additional distributed concentrated solar energy on the tube and an increase in energy absorption, storage capacity, and energy storage at higher temperature.

278a Production of Low Carbon Hydrogen Via Thermochemical Routes

Ahmed Salem, Ellie Bales, Aimaro Sanna, Raffaella Ocone

Much effort is currently being devoted to the production of green hydrogen (e.g. from renewables to electrolysis). However, oil majors, who are actively wishing to participate in the low carbon transition, need a starting point which is closer to their core business. Thus, the conversion of hydrocarbons to hydrogen provides a good entry point for many companies. Robust technologies to produce blue (low carbon) hydrogen at scale, and tools to determine how to deploy these optimally, alongside other low-carbon technologies at the local and national scales, are still in development.

The current work explores the suitability of biomass to produce hydrogen via pyrolysis. The work presents a computational model for analysing the performance of fast and intermediate pyrolysis in a fluidised bed pyroliser. The model investigates the impact of operating conditions and the physical and chemical properties of a specific biomass on the pyrolysis process. Through sensitivity analysis of the main operating parameters, including temperature, pressure and residence time, optimal conditions have been identified to maximise hydrogen yield whilst minimising the amount of undesirable by-products. The model is able to predict primary and secondary pyrolysis and the best operating conditions to optimise hydrogen yield. A rule is presented that makes the model generalisable to various biomass wastes.

278b Chemical Looping for Low Carbon Energy and Chemical Production

Luke Neal, Runxia Cai, Sherafghan Iftikhar, Vasudev Haribal, Fanxing Li

The high thermodynamic efficiency and inherent gas separation enabled by the chemical looping strategy offer significant opportunities for decarbonization of energy and chemical production. Historically, chemical looping research has focused on fossil fuel combustion, allowing for easy CO_2 capture and sequestration. In the past few years, we have studied chemical looping for a wide variety of chemical and energy production/conversion applications. For example, we have previously shown that chemical looping-based oxidative dehydrogenation can offer a 62 - 87% reduction in carbon emissions for the production of ethylene due to the high exergy efficiency and high single pass ethane conversions. In this work, we focus on two chemical looping pathways for decarbonization: Chemical Looping Air Separation (CLAS) and Chemical Looping H_2O/CO_2 Splitting. In CLAS, a mixed metal oxide is cyclically reduced in steam to give off gaseous oxygen and then oxidized in air to regenerate the oxide. We have shown that small-scale systems have the potential to produce oxygen at costs and energy demands comparable to or lower than large cryogenic air separation plants, greatly reducing parasitic losses for oxycombustion relative to pressure swing adsorption.

In chemical looping H_2O/CO_2 splitting, the mixed metal oxide catalyst is reduced by a fuel-rich gas stream and subsequently used to split either water into hydrogen or CO_2 into CO. This results in high-purity CO or H_2 without the use of extensive syngas conditioning and gas separation, allowing the production of low-cost hydrogen or conversion of captured CO_2 into a valuable chemical precursor. The fuel source for the reduction step can either be natural gas, which will produce high-quality syngas with a 2:1 H_2 to CO ratio, or an inexpensive, low-quality fuel gas stream from an industrial process. We present a detailed techno-economic analysis (TEA) and carbon emissions analysis of a system for the production of acetic acid from natural gas and CO_2. The potential integration of renewable energy to achieve negative net emissions of CO_2 is also discussed.

278d Catalytic Methane Decomposition: Effect of Carbon Yield on Particle Properties and Fluidization Characteristics

Dung Pham, Kang-Seok Go, Woohyun Kim, Hongjin Lee, Daewook Kim, Byungwook Hwang, Ho-Jung Ryu, Sang-goo Jeon, DongHyun Lee, Nam Sun Nho

Hydrogen production (Turquoise H_2) through the direct decomposition of methane has been receiving a lot of attention recently, as it produces carbon in solid form during the reaction process, eliminating the need for carbon dioxide capture and storage. Moreover, when transition metal catalysts are used, crystalline carbon can be produced, which can be utilized as a conductive material, potentially contributing to a reduction in the cost of hydrogen production. This technology typically utilizes fluidized bed reactors for the continuous production of hydrogen and carbon, where understanding the changes in particle properties and the resulting hydrodynamic characteristics within the reactor is very important for reactor design. This study aimed to investigate whether selective discharge of high-yield carbon particles is possible in the continuous methane catalytic decomposition reaction. For this purpose, particles with different carbon yields (based on a total weight change of particles/weight of fresh catalyst) around from 0 (Fresh Cat.) to 1000 wt.% were prepared in batch mode reaction. These were used to observe changes in particle properties (size, density, shape) according to carbon yield, and the mixing and separation characteristics according to fluidizing gas velocity changes were examined in a bubbling fluidized acrylic reactor with mixed carbon particles. Based on these results, an operational methodology for application to continuous processes was proposed.

(This work was supported by the Technology Innovation Program (20010853) funded By the Ministry of Trade, Industry & Energy (MOTIE, Korea) and was also conducted under the framework of the research and development program of the Korea Institute of Energy Research (C4-2404)).

278e Catalytic Fast Pyrolysis (CFP) Regenerator Model Development

Bruce Adkins, Mehrdad Shahnam, Yupeng Xu, Jordan Musser

Catalytic Fast Pyrolysis (CFP) is a promising technology for producing fungible transportation fuels from biomass. Just as in petroleum resid or vacuum gas oil (VGO) cracking in FCC units, carbonaceous coke deposits on the catalyst, and the heat from combusting the coke in a regenerator is used to drive the endothermic cracking reactions. However, biogenically derived coke is different from VGO and resid coke, and design of oxidative regenerators and regeneration processes must take these differences into account.

In coordination with the ChemCatBio (CCB), Advanced Catalyst Synthesis and Characterization (ACSC) and Catalyst Deactivation Mitigation (CDM) consortia funded by the DOE Bioenergy Technology Office (BETO), the Consortium for Computational Science and Chemistry (CCPC) has been investigating the kinetics of oxidation of coke generated on promoted and unpromoted ZSM-5 based CFP catalysts, and using these kinetics to build industrial scale reactor models. Two different ZSM-5 based catalysts have been investigated: a Geldart B (nominal 600-800 mm) coked to relatively high levels (12-15 wt% coke on catalyst), as would occur in a bubbling bed CFP reactor, and a Geldart A (FCC-type, nominal 80 mm) catalyst coked to much lower levels (< 2% coke on catalyst), as would occur in an FCC-type CFP reactor operating at typical catalyst circulation rates.

Kinetics model development was based on Temperature Programmed Oxidation (TPO) data generated at Pacific Northwest National Lab (PNNL), complimented by advanced microscopic characterization at Oak Ridge National Lab (ORNL). Fixed bed finite element computational models were used to assess mass and heat transfer limitations in the TPO experiments and incorporate learnings from characterization such as intrapellet coke deposition patterns. The fixed bed models were also used to test an ensemble of kinetic hypotheses including direct surface oxidation of C and H to form CO, CO_2 and H_2O, and homogeneous gas phase oxidation of CO to CO_2. A single best model with fifteen parameters was found to give an excellent fit to four key, very different TPO scans.

The kinetics model was translated into Barracuda® and MFiX-Exa, and used to simulate a 5 ton per day (TPD) demonstration-scale regenerator. Parametric studies of different gas-solids drag closures, air flow rates, solids inventories etc. were conducted and the process was optimized. Furthermore, an industrial-size catalyst regenerator for a 2500 TPD scale CFP unit was simulated with MFiX-Exa, a true exascale capable code, producing the first (to our knowledge) exascale regenerator simulation. With the MFiX-Exa model, different configurations of air grids and various operational modes were studied. The study has high value for addressing the potential problem of "afterburn" (CO combustion in freeboard) common to many FCC units.

278f Enhancing Carbon Capture Efficiencies in Natural Gas Power Plants through Magnetic Stabilization of Fluidized Beds

Ashin Sunny, Ishani Karki Kudva, Shekhar Shinde, Danwyn Aranha, Krutarth Pandit, Sudeshna Gun, Pinak Mohapatra, Dawei Wang, Andrew Tong, Liang-Shih Fan

Despite the increase of renewables for power generation, fossil fuels such as natural gas remain crucial in meeting our growing energy needs. However, fossil resources release significant amounts of CO_2, the primary greenhouse gas driving the pressing global issue of climate change. This underscores the urgent requirement to develop CO_2 capture technologies, including efficient CO_2 capture materials and reactor technologies for Natural Gas Combined Cycle (NGCC) power plants. The objective of this study is to improve the carbon capture efficiency of the NGCC process by employing a magnetic field-supported fluidized bed.

The work employs solid sorbent-based approaches for CO_2 capture, which are less energy-intensive and have wider operating temperature ranges compared to solvent-based or membrane-based methods. Specifically, solid sorbents based on potassium carbonate are not only cost-effective and possess high CO_2 sorption capacities but are also well-suited for post-combustion CO_2 capture due to their tolerance towards moisture in the flue gas. While several studies have focused on sorbent material design, there has been insufficient attention paid to reactor design for enhancing carbon capture efficiency. Traditionally, fixed bed reactors are used for CO_2 capture due to their operational simplicity. However, they suffer from poor heat and mass transfer characteristics, leading to high pressure drops when processing high flue gas flow rates. These limitations could be mitigated by fluidizing the bed, which improves heat transfer but may not necessarily enhance mass transfer due to bubble formation. Implementing a magnetically stabilized fluidized bed (MSFB) shows promise in increasing carbon capture efficiency by suppressing bubble formation, thereby enhancing gas-solid contact. This approach also reduces pressure drop across the bed, prevents sorbent carry-over at high flow rates, and minimizes sorbent attrition.

A magnetic bed setup was designed, utilizing a current-carrying coil cage simulated using COMSOL to generate a uniform magnetic field within the bed. The setup was fabricated, and hydrodynamic testing was conducted using sorbent material K_2CO_3 supported on Al_2O_3, impregnated with magnetic iron (Fe). Energy Dispersive Spectroscopy (EDS) confirmed the uniform dispersion of Fe on K_2CO_3-Al_2O_3. The magnetic susceptibility of the sorbent was determined using a superconducting quantum interference device (SQUID), quantifying the amount of magnetic field required to stabilize the bed. The hydrodynamic study investigated the effect of magnetic field strength, Fe loading, and gas velocities on bed pressure profiles. Results showed decreased pressure fluctuations when the bed entered the stabilized regime compared to when no magnetic field was applied, validating bubble suppression and enhancing gas-solid contact. Additionally, a decrease in pressure drop across the bed was observed when applying the magnetic field, contributing to potential cost savings on compressor operations in large-scale applications.

Following this, CO_2 capture from flue gas streams using K_2CO_3-Al_2O_3-Fe sorbent was evaluated across various reactor configurations - fixed bed, fluidized bed, and MSFB - at the same weight hourly space velocity (flue gas flow rate to weight of K_2CO_3). Characterization techniques such as X-Ray Diffraction (XRD) confirmed the formation of desired sorbent phases post-adsorption ($KHCO_3$) and post-regeneration (K_2CO_3). Performance across the different configurations was assessed based on carbon capture capacity and CO_2 removal fraction. The MSFB reactor demonstrated higher carbon capture capacities due to the improved gas-solid contact achieved by suppressing bubbles, as also observed in prior hydrodynamic testing. Compared to conventional fluidized beds, MSFBs also exhibited superior CO_2 removal fractions.

This study helps confirm the benefits of magnetic stabilization in fluidized beds, with potential applications in industrial processes requiring effective gas-solid contact. Insights gained from material and reactor design could also advance other CO_2 capture technologies vital for addressing climate change, such as direct air capture, which is currently a prominent area of research.

278g Design and Operation of a Novel Aluminum/Water Reactor to Produce H_2

Oscar Punch, Christopher Spitler, Christopher Boyce

Aluminum is an extremely energy dense material which has been proposed as a promising material for hydrogen storage technologies. The reaction of aluminum and water produces hydrogen gas and an aluminum oxyhydroxide byproduct. A key issue with prior aluminum/water reactors is that the aluminum oxyhydroxide byproduct coats the surface of unreacted aluminum particles, choking the flow of water to unreacted aluminum particles and limiting the reaction. In this work we present a novel aluminum/water reactor which utilizes fluidization and vibration to separate the unreacted aluminum and aluminum oxyhydroxide. We show that this novel design significantly increases the amount of aluminum which is reacted compared to previous aluminum/water reactor designs. This work is expected to significantly improve the feasibility of using aluminum as a hydrogen storage method.

278h Fluidized Bed Scale-up for Sustainability Challenges

Ray Cocco, Jia Wei Chew

The scaling up of fluidized beds has been purposefully pursued for more than 100 years. Yet, over that time, scale-up tools have not significantly changed. Data analysis is typically a standard analysis of variances (ANOVA) statistical exercise, perhaps reinforced with a design of experiment procedure. Flow sheeting and equipment design are based on institutional knowledge, albeit graphical user interface (GUI)-based process flow models make that job more manageable. Advanced models such as computational fluid dynamics (CFD) are used but often as a supplement and not a primary driver. As a result, the scale-up process for a fluidized bed can take more than ten years.

Fluidized beds remain at the forefront of the present time-critical sustainability challenges, e.g., carbon capture by particulate sorbents, methane-to-hydrogen, plastic-to-chemicals, etc.. In view of the exigency towards net-zero, today's scale-up efforts need to be accelerated, leveraging the advanced new tools that have become readily available. The problem is that such tools are often neglected, inadequately implemented, ineffectively resourced and/or poorly understood. This motivated the current effort, which is targeted at reviewing how scale-up tools have evolved over the years and the promising new tools, addressing some of the barriers of these tools in the design and scale-up of fluidized beds, as well as what can be done to circumvent these barriers. Moreover, we propose a new scale-up path leveraging the advanced tools to achieve timely implementation of the new green fluidized bed processes.

340a Development of a Liquid Bridge Model for Particle Agglomeration and Defluidization in Plastic Pyrolysis

Subhodeep Banerjee, Jean-Francois Dietiker, Mehrdad Shahnam

Pyrolysis is an important process for recycling and upscaling of plastics and municipal solid waste whereby the polymeric residues are reduced to basic raw materials that can be used in refineries or in the petrochemical industry. Fluidized bed reactors are commonly used for pyrolysis because they allow for the high value products to be extracted easily. Several recent studies have reported issues associated with agglomeration of the bed material when coated with fused (melted) plastic and consequent defluidization of the fluidized bed. The mechanism for the agglomeration is the cohesive forces arising from the formation of liquid bridges between the inert particles by the melted plastic. Previous research in the literature modeling the liquid bridge interactions using computational fluid dynamics (CFD) did not accurately account for the capillary forces between inhomogeneous particle pairs or the modified collision properties of coated particles. The implementation of the liquid bridge model also considered a fixed volume of liquid, which is inapplicable to a system with evolving liquid layer thickness as the melted plastic pyrolyzes. In this work, a novel liquid bridge model capability is implemented into the open-source CFD solver MFiX developed at the National Energy Technology Laboratory that explicitly models the mass, volume, and species of the liquid layer such that the evolution of the liquid bridge forces during pyrolysis can be accurately modeled.

The capillary force and normal and tangential viscous forces associated with the liquid bridge were developed from first principles and implemented into MFiX through a user-defined function. The implementation of the liquid bridge model without pyrolysis was quantitatively validated against the experimental pseudo-2D spouted bed of Tang et al. (*Ind. Eng. Chem. Res.*, 58, 2019) using glass beads coated with silicone oil and showed excellent match across a range of different liquid volume fractions and viscosities. The validated model was used to predict the agglomeration and consequent defluidization, identified by a decrease in the pressure drop, of sand coated with melted low-density polyethylene (LDPE) in inert conditions by updating the liquid viscosity and surface tension to representative values for LDPE based on literature. When the pyrolysis reaction was enabled, the key result was the recovery of the fluidization performance as measured by the pressure drop to the pre-agglomeration state as the LDPE layer pyrolyzed and the cohesive forces associated with the liquid bridge dissipated. The defluidization and subsequent recovery matched the evolution of pressure drop reported in literature. Finally, the sensitivity of the defluidization to the solids holdup, material size, and fluidizing velocity were investigated to optimize the operating envelop for minimal impact on the performance of the pyrolysis reactor.

340b Nuclear Techniques for Detailed Investigation of Cohesive Particle Flows

J Ruud van Ommen, Kaiqiao Wu, Rens Kamphorst, P. Christian van der Sande, Evert C. Wagner, Gabrie M.H. Meesters

There is an increasing interest in fine particles, a few micron or less than a micron in diameter. Such particles are often difficult to process because of their cohesive nature. Similar problems are encountered for wetted particles. This means, for example, that a regular fluidized bed does not provide sufficient mixing for such cohesive particles, but that additional methods such as stirring or vibration should be applied. This makes the particle flow even more complex than the already complicated behavior of a fluidized bed. In recent years, we have developed and deployed X-ray imaging and single-photon emission radioactive particle tracking to obtain detailed information about vibrated fluidized beds, stirred fluidized beds and horizontal stirred beds. In this work, we will discuss some examples of results obtained with these techniques.

The X-ray setup consists of a triplet of cone-beam sources (X-ray tubes) and three 1548-by-1524 pixel detector panels. It can be used to study systems up to 20 cm in diameter. For imaging, we often use only a single source and detector panel. The X-ray imaging reveals that for a fluidized bed of micro-silica, vibration reduces channeling, but does not significantly change the agglomerate size. Vertical vibration shows to be much more effective than elliptical vibration. However, it cannot prevent stratification in the bed. When using a stirrer to enhance the powder mixing, the design of the stirrer is of large influence: it should maximize the sweeping coverage and avoid the creating of gas pathways. When required, the three sources and panels can be used to make a full 3D tomographic reconstruction of a system under study.

While the X-ray imaging gives precise information about the density distribution, single-photon emission radioactive particle tracking can be used to obtain the trajectory of a representative particle. We use a polystyrene bead with an activated gold core – emitting gamma-rays – as the tracer particle. Three identical γ-radiation slit collimator detectors are placed equidistantly around the system of interest. We demonstrate for a horizontal stirred bed that increasing the agitator rotation speed and the fill level of the bed both lead to an increase in the solids circulation.

340c From Bouncing Space Probes to Toner Powders

Denis Schütz, Helena Weingrill, Natali Unterberger

Ever since the near-failure of the Philae lander of the Rosetta spacecraft mission in 2004, the behavior of Granular Materials (Bulk Solids) under conditions of low- or even microgravity has been an object of intense research scrutiny. Considering the planned future manned missions to the Moon and Mars, it is also likely to continue.

Due to the difficulty of actual experiments in microgravity, this work is so far mostly calculative and theoretical.

The applied stress towards a powder column is the most important parameter in the study of the flow (and non-flow) behavior of bulk solids, whereby the significance of the total applied force cannot be understated.

In the framework of the Mohr-Coulomb failure criterion applied normal stress is the key determinator of the behavior, often likened to temperature in soft matter rheology.

The key word here is applied stress, as the theory as well as the employed measurement techniques (Jenike or ring type shear cells) do not factor in artifacts that are contingent upon the powder's own weight.

In this fashion a common problem is shared between space physics and chemical engineering as this leads to artifacts, especially in the measurement of bulk solid of high density such as metal powders.

Utilizing sub-fluidized system to selectively counteract gravity in a fluidization column and measured by a modern Rheometer, we aim to reconcile experimental data with published calculative data obtained by various model systems of different density.

In contrast using applied vacuum under the powder column we intend to simulate applied stress within the same system and compare it to shear cell data.

340d Experimental Study and CFD-DEM Simulation of Vibrated Dry and Wet Spouted Beds

Christopher Spitler, Oscar Punch, Briana N. Cabrera, Christopher Boyce

Spouted beds are specialized fluidized beds which utilize rapid gas flow through a narrow inlet to circulate particulate material for chemical processes such as coating, drying, and gasification. The efficiency of spouted beds is highly dependent on the rate of particle recirculation, which is closely tied to the behavior of the spout region. Cohesive forces, such as those created by liquid, and small asymmetricities in gas fluidization can cause deflection of the spout which reduces the particle recirculation rate. In this work we present experiments and numerical simulations using CFD-DEM that show that spouted beds can be coupled with vertical vibration to reduce spout deflection in both dry and wet spouted beds. The effect of varying vibration strength, liquid viscosity, and background gas velocity are investigated to further understand the spout deflection mechanism.

340f Investigations on Dynamics of Agglomerates in a 2D Vibrated Fluidized Bed Using Pressure Signal and High-Speed Image Analysis

Liang Zhang, Haifeng Lu, Haifeng Liu, Xiaolei Guo

The transition of particle systems between solid-like and fluid-like regimes is driven by changes in density and in applied states. In this respect, mechanical vibrations in drive the unjamming transition have been proved to reduce the unconfined yield stress and cohesion, thus achieving effective particle fluidization. In the present work, the physical and flow properties of Micro-nano alumina powders were first characterized. A predictive model of inter-particle forces has been developed upon the Mohr's Circle and the Rumpf equation to analyze the complex particle interactions between different particle systems. We investigated the dynamic characteristics of agglomerates in a 2D vibrated bed and the mechanism under which the vibration energy excites the agglomerates motion and deformation in the vibration fluidization state using digital image processing and signal analysis. The effect of powder physical properties and vibration parameters (frequency, amplitude, etc.) on particle fluidization was analyzed.

419a An Industry-Focussed, AI-Driven Approach to the Calibration, Validation, and Optimisation of DEM and CFD-DEM Simulations

Kit Windows-Yule, Andrei-Leonard Nicusan, Dominik Werner, Ben Jenkins

Numerical methods such as the discrete element method (DEM), computational fluid dynamics (CFD), and multiphase particle-in-cell (MP-PIC) provide potentially powerful tools for the investigation and optimisation of diverse particulate and multiphase systems. However, without rigorous calibration and validation, their outputs may be inaccurate, or even entirely unphysical. The process of calibration can, however, prove highly time- and labour-intensive, and both calibration and validation are sorely lacking a clearly-defined 'Best Practice'. In this work, we present the results of a 5-year project funded by the International Fine Particle Research Institute, in which we worked with 12 industrial users of DEM in order to analyse the shortcomings of current methodologies and, based on our findings, develop a Best Practice. In this work, we detail the Best Practice determined, and how it may be integrated into a wider framework for the calibration, validation, and AI-driven optimisation of DEM.

The calibration methodology developed, termed Autonomous Characterisation and Calibration using Evolutionary Simulation (ACCES), uses evolutionary optimisation strategies coupled to DEM simulations to autonomously "evolve" toward a suitable set of DEM parameters to match a chosen set of characterisation data and/or experimental results. The outputs of ACCES are then validated against wider experimental data from the system of interest, here produced using positron emission particle tracking, a powerful technique facilitating high-resolution 3D imaging of particulate systems. A similar evolutionary strategy can then be used to optimise the simulated system according to a specific set of goals determined by the industrial end-user. These may be to increase throughput (and thus revenue) or, particularly important in the current age of climate crisis, improve sustainability through e.g. increased energy-efficiency or waste reduction.

Our work demonstrates, using industry-relevant case studies, a complete workflow for the calibration, validation and optimisation of diverse industrial and scientific systems, using a set of open-source, user-friendly tools.

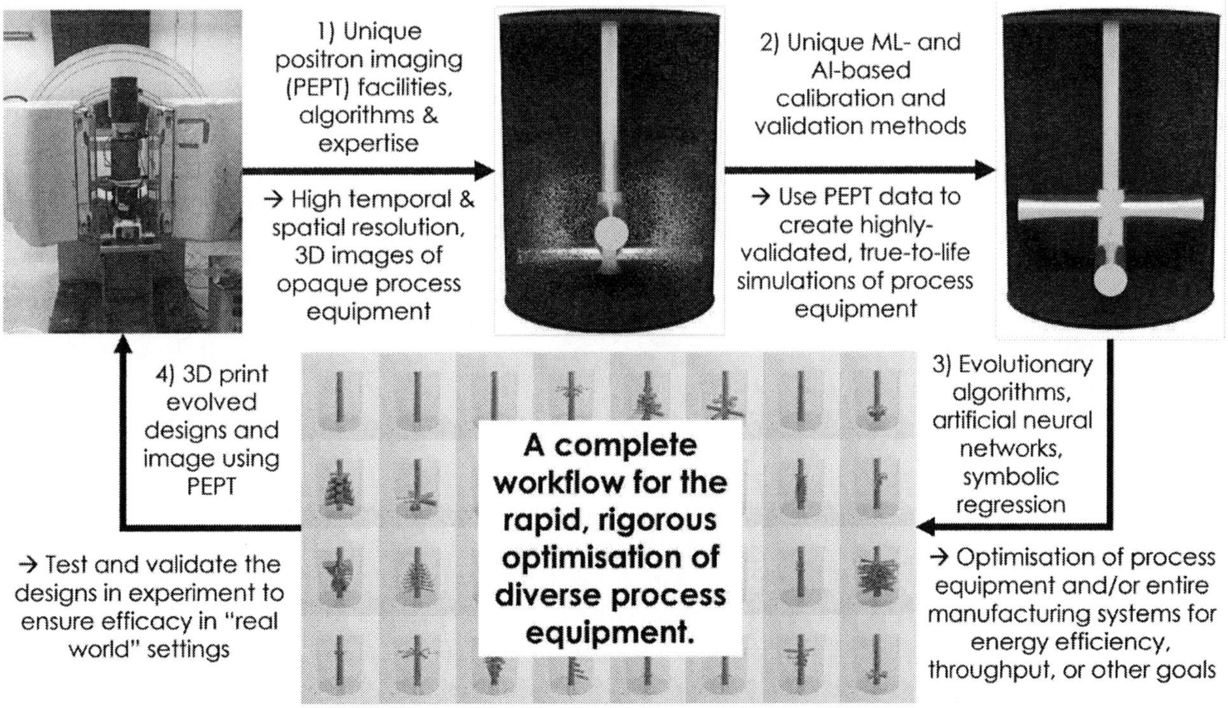

419b DEM Simulation of Particle Granulation in a Twin Screw Wet Granulator

Yu Wang, Pankaj Doshi, Eldin Wee Chuan Lim

Granulation is a method of particle size enlargement via agglomeration of primary particles to form larger granules. In a Twin Screw Wet Granulator (TSWG), a binder liquid is sprayed onto the surfaces of active pharmaceutical ingredient particles to promote the formation of larger granules with specific morphology, flowability, compressibility and other properties. The morphology of granules formed from a TSWG process depends on the material properties of the primary particles and operating conditions applied but mechanistic understanding of the granulation process remains incomplete even today. A large number of experiments are typically performed to determine the operating parameters required for a specific type of primary particles. The Discrete Element Method (DEM) has been applied extensively by researchers for simulations of various types of particulate processes. However, good agreements between DEM simulations of the TSWG and experimental measurements have not been demonstrated in the literature to date. In this study, the DEM was coupled with a capillary liquid bridge model to simulate the granulation process in a TSWG. Different liquid content (30%, 45% and 60%), screw speed (450 rpm, 675 rpm and 900 rpm) and solid throughput (5 kg/h, 12.5 kg/h and 25 kg/h) were applied to investigate the effects of these parameters on the residence time distribution of particles in the TSWG and size distribution of the granules formed. It was observed that agglomeration of particles occurred in the conveying elements of the TSWG and large granules were kneaded in the kneading elements to generate new granules. Granules were generated with different shapes and sizes at the outlet of the TSWG. The magnitudes of various forces, such as collision forces and capillary liquid bridge forces, were extracted for analyses and size distributions of granules formed at the outlet of the granulator were compared with experimental data reported in a previous study.

419c Effect of Triboelectric Charges on Powder Flow: DEM Simulation with Patchy Particles and Comparison with Experiments

Nicolas Preud'homme, Eric Opsomer, Aurelien Neveu, Geoffroy Lumay

Once granular materials flow, particles charge because of the triboelectric effect. When particles touch each other, charges are exchanged during contact whether they are made of the same material or not. Surprisingly, when different sizes of particles are mixed together, large particles tend to charge positively while small particles charge negatively. If the particles are relatively small (typically smaller than a millimeter), the electrostatic interaction between the particles becomes significant and leads to aggregation, intermittent flows or sticking on surfaces. Studying those effects is challenging as the mechanisms that govern the triboelectric effect are not fully understood yet.

DEM simulations - We show that the patch model (or mosaic model) is suitable to reproduce numerically the flow of triboelectrically charged granular materials in rotating drum geometry. We investigate the influence of charging on the cohesion of granular materials and highlight the relevant parameters related to the patch model that influence cohesion.

Experiments – A set of experiments were performed in the same geometry with GranuDrum instrument and the ability of the powders to charge was measured with GranuCharge instrument.

The experimental and simulation results are compared to shed new light on the mechanisms of the triboelectric effect as well as on how the charging of granular materials influences cohesion. Moreover, the links between these fundamental results and industrial processes will be discussed.

Reference:

Tribocharging of granular materials and influence on their flow, N. Preud'homme, G. Lumay, N. Vandewalle and E. Opsomer, Soft Matter **19**, 8911 (2023)

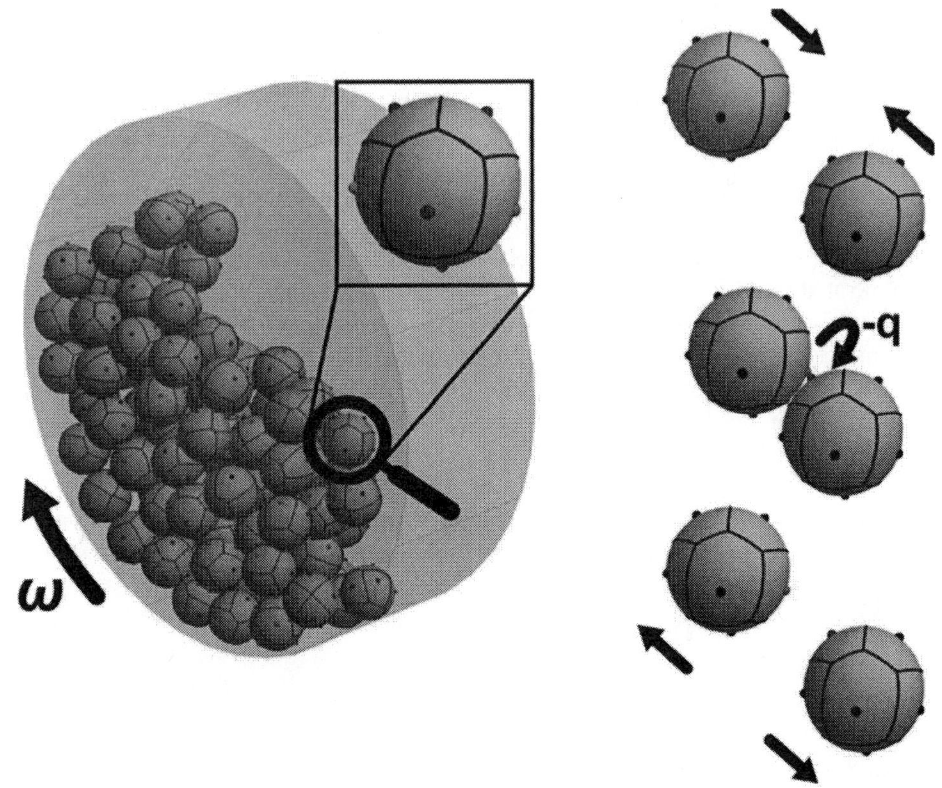

419d Complex Flow Patterns and Their Impact on the Performance of the Spiral Air Jet Mill

Kunal Sharma, Devang Khakhar

The focus of the present work is the analysis and optimization of an air jet mill, which is a versatile equipment used in many industries for fine-grinding powders to sizes less than around 10 microns. The efficiency of the grinding process in an air jet mill typically varies from 2% to 4%, depending on the design and operating parameters so even small efficiency improvements can result in significant energy savings [1]. Amongst many of the available machines, spiral air jet mills are attracting attention because of their advantages like absence of the moving parts and low-temperature rise during grinding, resulting in better temperature control with minimal maintenance. Spiral air jet mills boast an exciting ability: not only can they grind feed particles, but they can also classify the resulting material as a product. This distinctive feature truly sets these machines apart.

From the impetus of the literature, as a first step, we study the gas flow in the mill using Computational Fluid Dynamics (CFD) with OpenFOAM and actual machine behavior through experiments. The spin number is the ratio of tangential velocity to the radial velocity of the flow. A particle inside the jet mill experiences the drag toward the outlet in the radially inwards direction and centrifugal force in the radially outward direction. A balance of the two forces defines a cut size of the particle at which both the forces are equal and opposite in direction. As per the resultant equation, the cut size is inversely proportional to the spin number, assuming other variables to be constant. Thus, as the system is dilute of particles, getting the flow behavior of a single phase helps to predict the cut size variation.

Simulations with variable gas flow rate (Q), classifier height (ht), classifier radius (rt), rotation (RPM) of the peripheral surface, the chamber sizes (R), the chamber heights (H) of the machine, fluid kinematic viscosity and the machine scale-up. Only one parameter was varied at a time keeping others unchanged from a base case, for univariate analyses. We have found that the spin number decreases with a rise in input gas flow rate (Q) and with falling peripheral surface rotation (RPM), increases with higher chamber radial size and lower classifier radius, and the spin number is not affected significantly by chamber height and classifier height. As per the simulations, the machine spin number drops more steeply for higher fluid viscosities. On the experimental front, the runs with different chamber sizes were carried out. It was found that larger diameter chambers yield smaller product particles, the findings of which are by simulation results (higher spin number yielding smaller product cut sizes).

In base case simulations wherein the entire curved surface of the chamber is a rotational inlet, the airflow was observed at the steady state of the machine dynamics. It was observed that there is a presence of recirculation zones getting stronger with an increased spin number. This phenomenon led to the maximum velocity being concentrated near the walls of the chamber, potentially contributing to the misclassification occurring within the mill. Interestingly, the spin number exhibited a quadratic relationship with the size of the zone. The streamlines in laminar and turbulent conditions differ a little but the enlargement of the recirculation zone was evident in both regimes, while the velocity profiles are distinct. Flow patterns of simulations with jets are quite complex, because of the presence of many dead zones inside the flow area of the mill.

1. Mebtoul, M., Large, J., Guigon, P., 1996. High-velocity impact of particles on a target-an experimental study. Int. J. Miner. Process. 44, 77–91. https://doi.org/10.1016/0301-7516(95)00020-8.

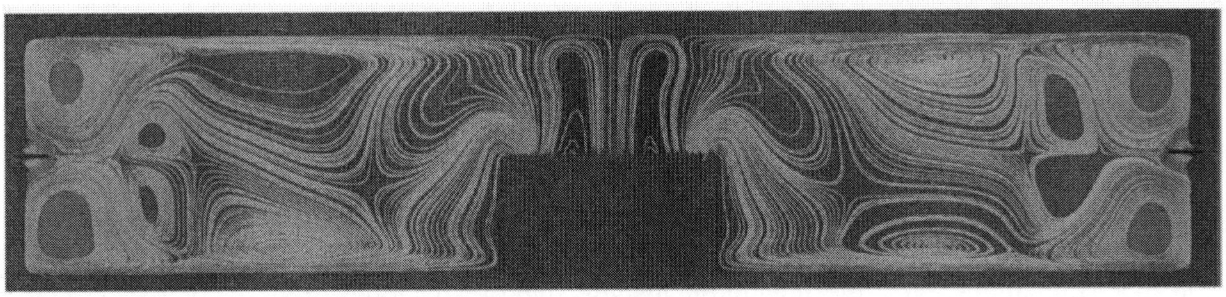

Flow streamlines for spiral air jet mill (with jets)

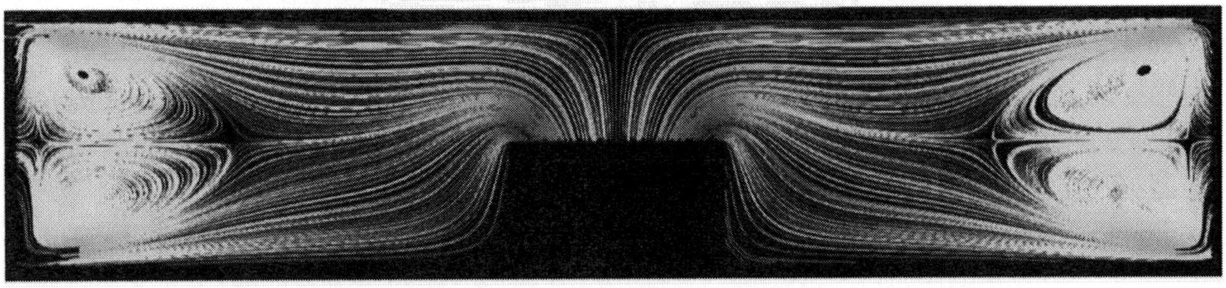

Flow streamlines for spiral air jet mill (without jets – Laminar)

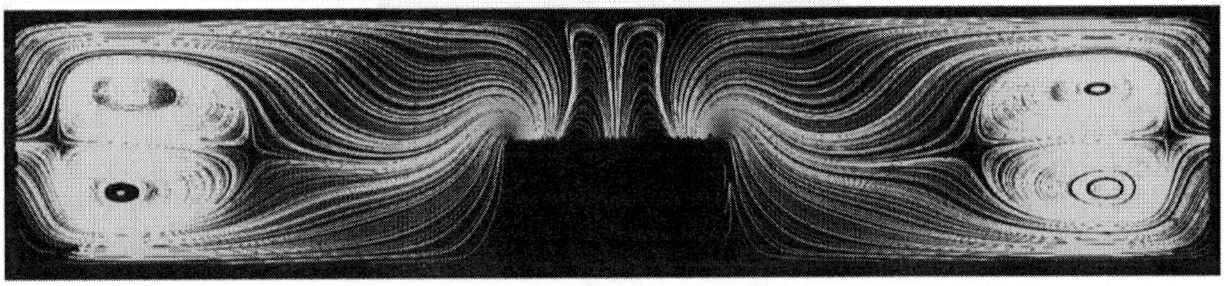

Flow streamlines for spiral air jet mill (without jets – Turbulent)

419f A CFD-DEM-Vof Model to Simulate Bubble-Liquid-Particle Three-Phase Flow

Facheng Gong, Yimei Chen

The simulation of interactions between bubbles, liquid, and particles is crucial for understanding the intricate mechanisms in three-phase systems. Numerous numerical models have been developed to tackle such simulations, each offering unique insights into the dynamics of these complex systems.

One widely used approach is the Volume of Fluid (VOF) method, first introduced by Hirt in 1981, which has been extensively applied to model liquid-gas multiphase flows. Additionally, Mulbah provided a comprehensive review of VOF methods for simulating bubble dynamics, shedding light on their utility and limitations. Moreover, various models, such as resolved CFD-DEM and unresolved CFD-DEM methods, have been developed to simulate particulate flow dynamics. These models offer valuable tools for investigating the behavior of particles in diverse environments.

Furthermore, researchers have begun to integrate multiple numerical models to capture the intricate interactions between liquids, bubbles, and particles. For instance, Ge et al. (2020) employed an unresolved VOF-DEM model to study particle dynamics within liquid slugs in a vertical gas-liquid-solid Taylor flow microreactor. Similarly, Shen et al. (2022) developed a resolved CFD-VOF-DEM model to investigate the interaction between gas-liquid two-phase fluids and irregularly shaped particles, showcasing the potential of combined approaches in capturing complex phenomena.

Despite the wealth of studies on three-phase flow, there remains a notable gap in research concerning wettability effects. Addressing this gap, this work proposes a novel CFD-VOF-DEM model that incorporates wettability interactions, offering a more comprehensive understanding of bubble-liquid-particle flows. By accounting for wettability effects, this model promises to provide deeper insights into the behavior of three-phase systems, facilitating advancements in various fields, from chemical engineering to environmental science.

419g Industrial Practice and CFD Investigation of a Novel Fast Fluidized Bed for Fluid Catalytic Cracking

Wenming Liu, Jianhong Gong, Jingxiao Wang, Jinquan Zhu

Fluid Catalytic cracking (FCC) is a crucial technology for producing propylene from heavy oil. The riser is widely used in FCC technology due to its high gas-solid flux and short particle residence time. However, the low gas-solid contact efficiency of the riser is prone to thermal cracking reactions, resulting in producing more dry gas and coke, which limits the improvement of propylene selectivity. Based on the idea of enhancing gas-solid contact efficiency, a novel fast fluidized bed was proposed to improve propylene yield, and the industrial practice and CFD investigation were performed. The industrial practice results showed that in comparison with the typical riser technology, the yield of ethylene and propylene increased by 0.5 and 2.56 percentage points respectively, and the gasoline research octane number (RON) increased by 0.4. Besides, dry gas and coke decreased. To analyze differences between the conventional riser and the novel fast fluidized bed, a structure-based transfer model coupled with 3D CFD was established. The simulation results showed that the model predicted the gas-solid flow behavior more accurately than the traditional model. The computed solids concentration and particle residence time in the novel fast fluidized bed were higher than those in the riser, indicating that the novel fast fluidized bed exhibited higher gas-solid contact efficiency. Therefore, the results suggest that the novel fast fluidized bed has promising potential for propylene production.

419h CFD Modeling of Binary Mixing of Particles in Fluidized Bubbling Bed Reactors

Kuochen Tsai, Guoqiang Yang, Piet Huizenga

Biomass processing in fluidized bubbling bed is getting attraction for renewable liquid fuel productions in recent years [1,2]. However, there are still many difficulties and uncertainties regarding the process yield, stability and their scale-up. Instead of biomass pyrolysis or gasification, we took a different route by synthesizing liquid hydrocarbons directly. In this presentation, we will discuss how some of these issues were addressed using experimental data and computational fluid dynamics (CFD) modeling, especially regarding how the biomass and catalyst mixes in a bubbling bed with hydrogen as the fluidization gas.

The binary mixing process is designed to extract oxygenate vapor from biomass (non-food organic waste) and react it with hydrogen in a bubbling catalyst bed. The process requires proper dispersion of biomass with catalyst particles, in which it generates heat through deoxygenation reaction and produces hydrocarbon vapor under high temperature and high pressure. The yield depends critically on the degree of mixing, and the efficient generation of biomass oxygenate vapor at the early stage. If not properly designed, it can significantly impact reactor performance in many ways. The use of CFD enabled us to predict and identify possible issues early on. Interactions with lab experiments and the demonstration reactive unit further enhance our understanding of the interplays between experiments and CFD models. Additionally, the details in CFD allow the quantification of many scale-up rules, including solid flux, binary mixing and catalyst loss, etc.

However, here the discussion will be limited to the CFD modeling and quantification of the binary mixing process and how it impacts the overall reactor design. The CFD model used is based on multiphase-particle-in-cell (MP-PIC) approach [3] due to its efficiency in modeling particle-particle interactions and a commercial code was adopted [4]. The model was first tested against published work for binary mixing of particle and later was further fine-tuned to mimic the particle segregation time-scale measured in experiments. This establishes the fundamental mixing correlations between the particles of interest. Due to the computational costs of chemical reaction calculations, a simplified thermal mixing method was used to mimic the heat produced from the exothermic chemical reaction by assuming a high catalyst heat capacity that retains heat. The room temperature fed biomass particles provide the cold quenching experienced by the catalysts. The biomass particles can track the temperature values as they travel through the reactor and generate a temperature mixing history to validate required particle heating performance. When used with a demonstration unit data, it can show the necessary performance curve needed for the scale-up reactor to follow.

The experience we obtained from the interactions between CFD modeling and experiments are invaluable and provided many practical and useful lessons regarding how both can gain benefits from each other, especially clear and open communications and well defined processes are critical in generating positive feedbacks and creating productive results.

References:

[1] Troiano, et al., "Fluidized bed pyrolysis of biomass: A model-based assessment of the relevance of heterogeneous secondary reactions and char loading," Energy Fuels, 36, 17, 9660-9671, 2022.

[2] Beheshti, et al., "Process simulation of biomass gasification in a bubbling fluidized bed reactor,"

Energy Conversions and management, 94, 345-352, 2015.

[3] Peter J. O'Rourke and Dale M. Snider, 'A newblended accelerationmodel for the particle contact forces induced by an interstitial fluid in dense particle/fluid flows,' Power Technology, 256, 39-51, 2014.

[4] Barracuda Virtual Reactor 23.0.1, CPFD-Software Inc., 2023.

279a Dynamics of Molecular Collisions in Air and Its Mean Free Path

Dimitrios Tsalikis, Vlasios Mavrantzas, Sotiris E. Pratsinis

The kinetic theory of gases (Maxwell, 1860; Jeans, 1967) offers a remarkable description of molecular collisions in gases with its hard-sphere model and under the assumption of collisions being purely elastic. In practical applications, however, such as in the aerosol synthesis of small nanoparticles at high temperatures and/or low pressures, one must accurately account for gas-particle collisions as asymptotic (self-preserving) particle size distributions and (fractal-like) structure have not been attained to simplify aerosol dynamics (Pratsinis, 2010). Then, the classic hard-sphere model might be insufficient to describe them. Motivated by such a need, we compare the kinetic theory with detailed molecular dynamics (MD) simulations for the dynamics of molecular collisions in air accounting, however, for oxygen and nitrogen as diatomic molecules, for the first time to the best of our knowledge.

The MD simulations were conducted in the NVE ensemble using simulation cells containing 21% mole O_2 and 79% mole N_2 with a fully atomistic (FA) model (Zambrano *et al.,* 2014). The simulations were carried out at 300 K and 1 atm to validate our methodology with textbook results for density and transport properties. A distance criterion was set to identify collisions based on the separation between colliding molecules, as in kinetic theory. The force field employed was validated by comparing the MD-predicted air density (1.18 kg/m^3) to the experimental for dry air (1.18 kg/m^3). A second validation test involved a comparison of the MD-predicted values for the diffusion coefficient and viscosity with available measured values. Again, a good agreement was found.

Treating O_2 and N_2 as hard spheres, their collision densities by the MD simulations were also in agreement with their densities from kinetic theory (Jeans, 1967). When, however, these densities were calculated with the FA model, they were distinctly above those of the kinetic theory. This is attributed to the attractive part of the Lennard-Jones (LJ) interactions and their diatomic, thus more voluminous, shape. To this, simulations were repeated with O_2 and N_2 considered as LJ spheres. Their collision densities fell between those from the kinetic theory and the FA model.

A detailed analysis of the extracted MD trajectories revealed that molecular collisions by the FA model involve strong interactions of the colliding species. There were instances where three or four molecules were involved in the same collision. Moreover, frequently, colliding molecules split from each other but soon returned to collide again and again without having interacted with any other molecule in between. Such collisions are termed spurious. Systematically, spurious collisions were identified and removed in the analysis to avoid overcounting collision events. These complex collision patterns were detected by the LJ model (Figure 1a) and the FA model (Figure 1b), but they were never observed by the HS model.

A direct result of the enhanced interactions between air molecules when these are treated as diatomic molecules is that their mean free path (MFP) (calculated both by the collision densities and by direct averaging over the probability distribution of free paths) comes out to be considerably smaller than that from the classic HS potential. The new value of the MFP of air, λ, reported here from our all-atom MD simulations is 38.5 ± 1 nm (Tsalikis et al., 2023), almost 43% smaller than the currently known and widely used value of 67.3 nm at 300 K and 1 atm (Jennings, 1998). The MD-extracted λ, is quite robust to the employed MD force field as long as it accurately matches the experimentally determined macroscopic properties of air (density, viscosity, and diffusivity).

References

J. Jeans, *An introduction to the kinetic theory of gases*, (Cambridge University Press, Cambridge, 1967).

S.G. Jennings, The mean free path in air, *J. Aerosol Sci.* **19** (1998) 159-166, doi.org/ 10.1016/0021-8502(88)90219-4.

J.C. Maxwell, Illustrations of the dynamical theory of gases-part I. On the motions and collisions of perfectly elastic spheres, *The London, Edinburgh, and Dublin Philos. Mag. and J. of Science* **19** (1860) 19-32, doi.org/ 10.1080/14786446008642818.

S.E. Pratsinis, Aerosol-based technologies in nanoscale manufacturing: From functional materials to devices through core chemical engineering, AIChE J. 56 (2010) 3028-3035, doi.org/ 10.1002/aic.12478.

D.G. Tsalikis, V.G. Mavrantzas, S.E. Pratsinis, Dynamics of molecular collisions in air and its mean free path, *Phys. Fluids.* **35**, (2023) 097131, doi.org/10.1063/5.0166283.

H. Zambrano, J.H. Walther, R. Jaffe, Molecular dynamics simulations of water on a hydrophilic silica surface at high air pressures. *J. Mol. Liq.* **198** (2014) 107-113, doi.org/ 10.1016/j.molliq.2014.06.003.

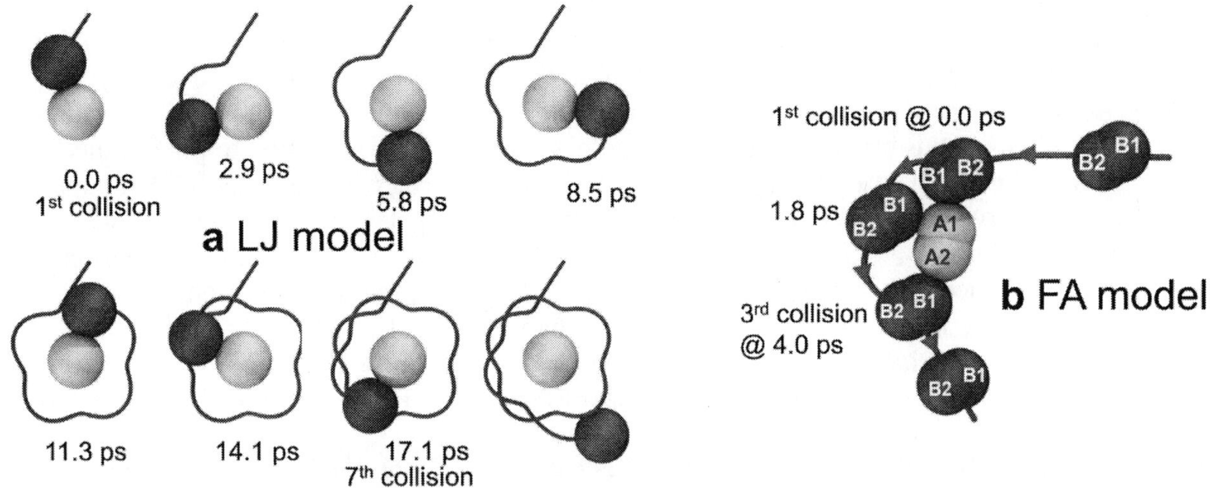

Figure 1. Spurious collisions (successive ones between the same molecules) in relative coordinates during a single otherwise collision by: (a) The LJ model with the pair of colliding molecules (depicted with green and blue beads) experiencing seven spurious collisions during a collision event. (b) The FA model (multimedia view) with the pair of colliding molecules experiencing three successive (spurious) collisions. For clarity, the rotational motion of the reference molecule (green) has been suppressed here. In both models, the line shows the trajectory of the moving molecule. These successive collisions are not independent events but part of a single collision.

279b Predictive Models for the Formation and Rigorous Design of Nanoparticles

Wolfgang Peukert, Nabi Traore, Lukas Pflug

Today, nanoparticle technology moves forward from careful control of "only" particle size to systems with higher complexity, i.e. shape, surface, structure and composition, both on the level of single particles and particle ensembles. Based on unifying principles for the production of particulate products, which are widely applicable to many different kinds of products, we discuss options for scalable production. Despite the amazing progress in nanomaterial synthesis, the transfer to industrial application is much less developed and must still be considered as a limiting challenge.

We demonstrate how to scale-up production of nanoparticles by careful consideration of the driving forces for particle formation in batch and continuous systems. Our modelling approach for mixing-controlled systems is based on direct numerical simulation of the fluid flow coupled to population balances and reaction equilibria of multicomponent systems [1]. We show how to predict the full particle size distribution in case of anti-solvent precipitation of drug nanoparticles. Widely applicable scaling laws are derived and applied to several other materials [2]. The complex formation mechanism of the reaction-controlled formation of Au, Ag and Au@Ag alloy nanoparticles is elucidated and modelled by a 2D population balance model predicting the evolution of size, composition and the resulting optical properties [3]. Beyond synthesis, whole process chains must be considered. Examples of rigorous, knowledge-based product design related to optical properties as studied in the collaborative research center "Design of Particulate Products" will be highlighted.

Funding by DFG of the Collaborative Research Center 1411 "Design of Particulate Products" is greatfully acknowledged (https://www.crc1411.research.fau.eu/).

References

[1] Schikarski T., Avila M., Trzenschiok H., Güldenpfennig A., Peukert W., Quantitative modeling of precipitation processes, Chem Eng. J. 444 (2022) 136195

[2] Schikarski T., Avila M., Wolfgang Peukert W., En route towards a comprehensive dimensionless representation of precipitation processes, Chem. Eng. J. 428 (2022) 131984

[3] Traoré N.E., Schikarski T., Körner A., Cardenas Lopez P., Hartmann L., Fritsch B., Walter J., Hutzler A., Pflug L., Peukert W., Mechanistic insights into silver-gold nanoalloy formation by two-dimensional population balance modeling, Chemical Engineering Journal 484 (2024) 149429

279c Particle Size and Shape Analysis and Reconstruction Using Two-Point Correlation Function for a Nanoparticle System

Puneet Koli, Rajdip Bandyopadhyaya, Abhijit Chatterjee

Nanoparticles play a key role in the fields of biomedicine, catalysis, electronics and environmental science. Crucial to the optimization and control of nanoparticle during synthesis is the material structure-property aspects which finally dictates their specific roles/function. In particular, the nanoparticle size and shape play an important role in determining the physical properties of the nanoparticles in various industrial applications. Although, there are many experimental and numerical techniques to calculate the nanoscale structure properties, there is a lack of techniques that can adequately calculate size and shape parameters for complex anisotropic nanoparticle shapes. This is mainly because the statistical aspects of the morphology need to be considered. In this talk, I present an all-encompassing algorithm that is based on the two-point correlation (2PC) functions which can directly provide the particle size, shape, and interparticle spacing as well as their distributions. Only a one-time effort is required to calculate the 2PC. Application of the technique to anisotropic nanoparticle systems, particles agglomerate, time-dependent shape evolution, 3D porous structures obtained from experimental characterization (microscopy imaging and tomography techniques) and computer simulations shall be demonstrated. The 2PC can be used to reconstruct a nanoparticle system which can serve as input to simulations. We believe that such a powerful tool can be beneficial for control and optimization of nanoparticle systems.

279d Molecular Dynamics Simulations and Machine Learning Assisted Size and Temperature Dependent Shape Analysis of Metal Nanoparticles

Huaizhong Zhang, Kristen Fichthorn

Metal nanocrystals are useful in a host of applications, ranging from catalysis to plasmonics to electronics to various personal and health applications, and many studies show there are optimal nanocrystal sizes and morphologies for these applications. In the interest of reliably synthesizing optimal nanocrystals, it is important to understand the shape trajectories nanocrystals follow as they grow. In a self-seeding synthesis, the small (likely single-nm) nanocrystal seeds that form after nucleation are fluxional and we expect them to assume an equilibrium shape distribution as they grow. As the nanocrystals grow larger, they may become locked into kinetic shapes. It is important to understand the thermodynamic-to-kinetic shape transition to design processing routes by which nanocrystals with specific sizes and shapes could be synthesized. Such an understanding hinges on resolving the thermodynamics of nanocrystal shapes.

In this study, we conducted Parallel-Tempering Molecular Dynamics (PTMD) simulations to investigate the temperature- and size-dependent equilibrium shapes for both mono-metallic (Cu and Ag) and bi-metallic (AgCu) nanocrystals in the 100- to 200-atom size range. Additionally, we employed Common Neighbor Analysis (CNA) and unsupervised machine learning methods to classify the shapes for each nanoparticle into different structural motifs.

Our PTMD studies of Cu nanocrystals reveal a preference for shapes intermediate between decahedra and icosahedra, which we designate as Dh-Ih, due to vibrational entropy effects. Analyzing temperature-dependent shape distributions across various sizes highlights the significant role of entropy in shaping Cu nanocrystals, indicating that studies solely focused on determining minimum-energy shapes may not accurately predict experimental shapes at varied temperatures. Additionally, significant shape changes are observed with changes in size, including alterations occurring with just a single atom. Our simulations of Ag nanocrystals also exhibit the Dh-Ih. Entropy-related phenomena are also prominent in Ag nanocrystals. For instance, pure FCC nanocrystals are less common than those containing stacking faults, and minimum potential energy shapes are not always favored across all temperatures.

In the study of AgCu bi-metallic nanocrystals, we frequently encountered nanocrystals with distinctive combinations of CNA signatures that deviate from the commonly applied empirical classification guidelines. Thus, we opted to employ machine learning methods to classify these shapes in a more robust and automatic manner. By combining the descriptors of nanoparticle structures based on their atomic environment and composition with unsupervised learning methods such as K-Means and Gaussian mixture method, we can distinguish between different and more detailed structural motifs. Overall, our study underscores the intricate interplay between temperature, size, and chemical identity in shaping the morphology of metallic nanocrystals. The insights gained from this study hold potential for informing the development of processing strategies aimed at achieving selective nanoparticle shapes.

279f Data Intensification in Flow for Accelerated Synthesis Space Mapping of Inorganic Nanomaterials

Fernando Delgado Licona, Hannah Dickerson, Abdulrahman Alsaiari, Milad Abolhasani

The ability to accelerate the exploration of synthesis–property relationships of functional materials could significantly boost the rate at which new materials are discovered, optimized, and manufactured. Self-driving labs (SDLs) have risen as a sustainable tool to reduce the time-to-solution in chemical and materials sciences by integrating lab automation, robotics, and artificial intelligence (AI) to enable closed-loop, self-guided experiments based on the human-defined scientific problem [1]. The SDLs' developments over the past few years have been driven by process intensification (PI) principles, to achieve faster, safer, and more efficient processes toward Self-Driving Fluidic Labs (SDLs) that leverage flow reactors with in-situ characterization techniques.

Despite the success of SDFLs in reducing the time-to-solution in chemical and materials sciences, they suffer from prolonged waiting times, focusing only on steady-state measurements and ignoring the transient measurements. However, if harnessed properly, the transient information of flow chemistry platforms can provide valuable insights into the underlying process mechanisms and could be used to expedite experimentation, provide actionable feedback to process controllers and enrich the AI agent of the SDFL [2,3]. In this work, we present dynamic flow experiments as a data-rich strategy for rapid quantum dot (QD) synthesis-parameter—material-property mappings at least 100× faster than steady-state flow chemistry strategies. Dynamic flow experiments take advantage of *in-situ* characterization techniques capable of generating time-series data from sudden or controlled changes in one or more process parameters. Through mapping the instantaneous to steady-state equivalent residence time, we demonstrate the unique data intensification capability of continuous flow reactors (>100× experimental data throughput) while reducing the total chemical consumption (~3) in a shorter time when compared to traditional steady-state flow experiments in the same system.

We use Cadmium Selenide (CdSe) QDs as a proof-of-concept case study for accelerated material parameter space mapping. This data-intensified approach leads to a considerable improvement in terms sustainability of SDFLs. Furthermore, the resulting Big experimental data generated from this data intensification strategy (>10,000 experimental data/day) highlights a sustainable pathway to generate a digital-twin representation of the reactive system for subsequent autonomous closed-loop experimentation.

References:

[1] M. Abolhasani, E. Kumacheva, The rise of self-driving labs in chemical and materials sciences, Nat. Synth. 2 (2023) 483–492. https://doi.org/10.1038/s44160-022-00231-0.

[2] F. Florit, A.M.K. Nambiar, C.P. Breen, T.F. Jamison, K.F. Jensen, Design of dynamic trajectories for efficient and data-rich exploration of flow reaction design spaces, React. Chem. Eng. 6 (2021) 2306–2314. https://doi.org/10.1039/D1RE00350J.

[3] K.C. Aroh, K.F. Jensen, Efficient kinetic experiments in continuous flow microreactors, React. Chem. Eng. 3 (2018) 94–101. https://doi.org/10.1039/C7RE00163K.

279g Modeling of Silver Nanoparticle Synthesis and Trapping at the Interface of an Aqueous Two-Phase Systems

Jagadeesh Korukonda, Subramaniam Pushpavanam

This study focuses on modelling the synthesis of silver nanoparticles (AgNPs) within microchannels using an aqueous two-phase system (ATPS). ATPS offers a biocompatible alternative to traditional organic-aqueous systems. The behaviour of the system is described using Computational Fluid Dynamics (CFD) and Population Balance Modeling (PBM) techniques. The flow patterns and reactant distributions are analysed using CFD, while the PBM accounts for particle nucleation, growth, and other interactions that determine the AgNP size distribution. The modified Finke-Watzky kinetic model is adpted to predict the AgNP formation accurately, including nucleation and growth kinetics. The study predicts the size and distribution of AgNPs in ATPS from the model and provides the conditions for nanoparticle trapping at the liquid-liquid interface in ATPS. This study provides crucial insights into the control and manipulation of nanoparticle properties during synthesis. The proposed model helps optimise AgNP synthesis efficiency for two-phase systems within microchannels.

279h A Simulation Framework for Nanodroplet Breakup in Top-Down Nanoparticle Production

Nick Morse, Johan Remmelgas, Johannes Khinast

Introduction

Nanoparticles have shown great promise as a pharmaceutical carrier, evidenced most strongly by the success of lipid nanoparticles as carriers for mRNA vaccines during the COVID-19 pandemic. However, the fundamental processes that govern nanoparticle formation are still not well understood and particle size predictions are generally reliant on empiricism. Numerical simulations offer the potential to inform a variety of nanoparticle production methods, such as fragmentation of particle suspensions in impinging jets, turbulent mixing for flash nanoprecipitation and nanoparticle self-assembly, and nanoemulsion formation. However, simulations are impeded by the vast separation of length and time scales between the pharmaceutical processing scale and the nanoparticle scale: the nanoscale is around one million times smaller than the production scale, and the disparity in timescales one million times greater still. Accordingly, the resolution of all scales in a single simulation is computationally infeasible.

In order to overcome this computational hurdle, a coupling framework between simulations of the macroscale production device and mesoscale behavior is required. Considering nanoemulsion formation, one may take advantage of the fact that nanodroplet sizes are smaller than the smallest turbulent (Kolmogorov) scales of the flow[1], which results in a low droplet Reynolds number based on the shear rate, diameter, and carrier fluid kinematic viscosity. This, combined with the small Stokes number of the droplets, permits a coupled macroscale-mesoscale simulation framework.

Methodology

The simulation framework consists of two parts: on the macroscale, the nanoparticle production device is simulated using a traditional finite volume method computational fluid dynamics (CFD) solver. Massless tracers are advected through the flow to record the strain rate histories along the Lagrangian trajectories of the droplets. Since the strain rates are dominated by the smallest turbulent scales, computationally-intensive direct numerical simulations (DNS) or large-eddy simulations (LES), combined with a strain rate reconstruction technique[2], are required.

These Lagrangian strain histories are used as input to an ensemble of auxiliary simulations of the droplet deformation. The droplet shape deformation is calculated using a boundary element method (BEM) to resolve the viscously-dominated flow around the droplets. In the BEM, only the droplet interface is tracked, reducing the computational expense for each calculation. In this framework, the effect of differing droplet/carrier viscosities, surface tensions, and surfactants are readily investigated[3,4], and the breakup of droplets can be attributed to specific regions of the mixing device.

Results

We discuss results from both the macroscopic and mesoscopic simulations. On the macroscale, the flow-fields in nanoparticle production devices are investigated, including the insights simulations provide into the physics of the most critical regions in these devices. On the nanodroplet scale, we present results of droplet deformation and discuss the incorporation more complex nanoscale effects in the simulation framework.

Literature

1. Gupta, A., Eral, H. B., Hatton, T. A., & Doyle, P. S. (2016). Controlling and predicting droplet size of nanoemulsions: Scaling relations with experimental validation. *Soft Matter*, *12*(5).

2. Johnson, P. L., & Meneveau, C. (2018). Predicting viscous-range velocity gradient dynamics in large-eddy simulations of turbulence. *Journal of Fluid Mechanics*, *837*, 80–114.

3. Pozrikidis, C. (1992). *Boundary integral and singularity methods for linearized viscous flow*. Cambridge University Press.

4. Cristini, V., Blawzdziewicz, J., Loewenberg, M., & Collins, L. R. (2003). Breakup in stochastic Stokes flows: Sub-Kolmogorov drops in isotropic turbulence. *Journal of Fluid Mechanics*, *492*, 231–250.

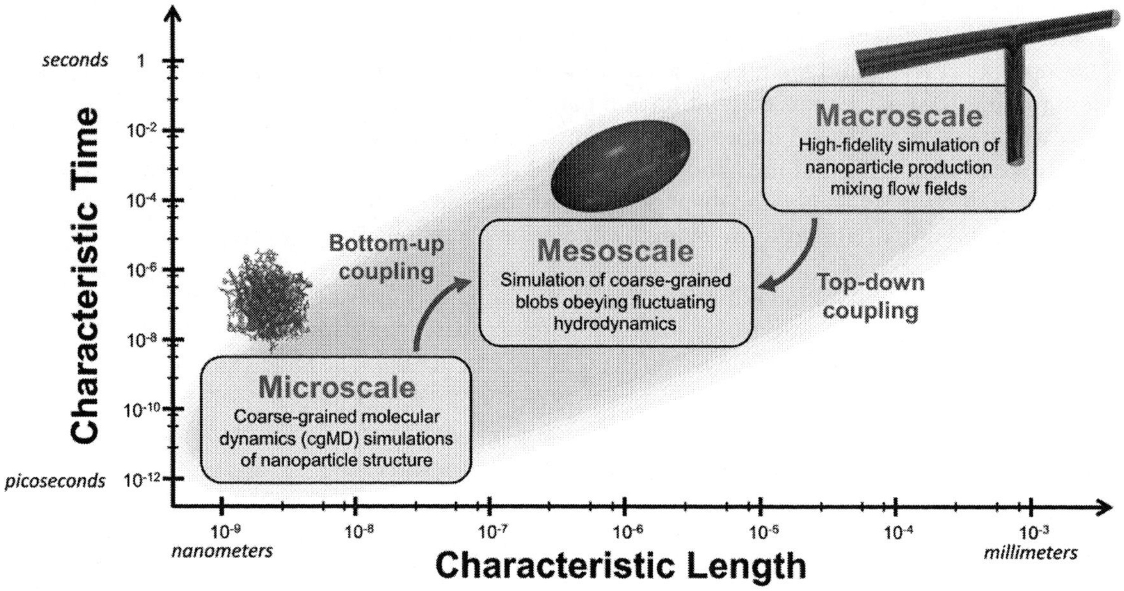

AUTHOR INDEX